# Keep Chickens!

# Keep Chickens!

## Tending Small Flocks in Cities, Suburbs, and Other Small Spaces

■ ■ ■ ■ ■ ■ ■ ■ ■

BARBARA KILARSKI

Storey Publishing

*The mission of Storey Publishing is to serve our customers
by publishing practical information that encourages
personal independence in harmony with the environment.*

Edited by Nancy W. Ringer
Art direction and cover design by Meredith Maker
Text design and production by Eugenie S. Delaney
Cover illustrations by © Lisa Adams
Interior photographs: Cara Smith: 119; Alan Stanford: 120; Diana Andersen: 121;
© Joel Sartore/Grant Heilman Photography: 122; © Larry Lefever/Grant
Heilman Photography: 123; © Howard Buffett/Grant Heilman Photography:
124; © John Colwell/Grant Heilman Photography: 125; © Bill Adams: 126;
© Mark Turner/Midweststock: 127; Steven Ellson: 128; © 2003 Geoff Hansen:
129; © Harley Soltes/Seattle Times: top, 130; W. Stephen Keel, Eggantic
Industries: bottom, 130; © Forsham Cottage Arks: top, 131; © Harley
Soltes/Seattle Times: top and bottom, 131; © Harley Soltes/Seattle Times: top,
132; Brighton West: bottom, 132; © Harley Soltes/Seattle Times: 133; Barbara
Kilarski: 134.
Illustrations by Elayne Sears
Indexed by Eileen M. Clawson

Text copyright © 2003 by Barbara Kilarski

The information in this book is true and complete to the best of our knowledge.
All recommendations are made without guarantee on the part of the author or
Storey Publishing. The author and publisher disclaim any liability in connection
with the use of this information. For additional information please contact Storey
Books, 210 MASS MoCA Way, North Adams, MA 01247.

Storey books are available for special premium and promotional uses and for
customized editions. For further information, please call Storey's Custom
Publishing Department at 1-800-793-9396.

Printed in the United States by Versa Press
10  9  8  7  6  5  4  3  2  1

**Library of Congress Cataloging-in-Publication Data**

Kilarski, Barbara.
   Keep chickens! Tending small flocks in cities, suburbs,
and other small spaces / Barbara Kilarski.
      p. cm.
      ISBN 1-58017-491-4 (alk. paper)
   1. Chickens. I. Title.
      SF487 .K5616 2003
      636.5—dc21

                                                2002154373

This book is dedicated to my human flock:
George and Yolanda, the best parents a chick could ask
for, and my sister Annette, a chick's best friend.
■ ■ ■ ■ ■ ■ ■

# Contents

■ ■ ■ ■ ■

# Acknowledgments
■■■■■

 PECIAL THANKS to Kim Kittinger, whose perpetual and enthusiastic daydreaming made it possible for me to pursue my own dreams.

# Introduction
■ ■ ■ ■ ■

T IS THE YEAR OF THE CHICKEN. Whether prompted by a practical attraction to the daily fresh eggs or by a fanciful desire to import a little bit of country charm in our urban lives, over the past two years city and suburban households have adopted more chickens than ever. Seattle. Portland. Los Angeles. New York. City chickens are hatching in urban areas at break-egg speed.

Why the sudden population boom in town and country chickens? Perhaps because chickens hearken back to simpler times when most American households included chickens. A time when most of our daily foods were brought in from the backyard, not shipped over thousands of miles and dozens of days to our dinner plates. Keeping chickens gives people a chance to slow down and smell the flowers — flowers that grow larger and brighter each year when composted with homegrown chicken poop.

Chickens have awakened the carpenters and craftsmen apparently latent in modern civilization. City chickens don't live in scrap wood shacks; they live in palatial hen habitats and estates built especially for them. Farmer wanna-bes in urban neighborhoods are investing considerable time and money in coop design and architecture. Maybe you can't afford to custom build a Swiss chalet or log cabin–style home for your family, but you can build one for your chickens instead!

Raising chickens satisfies an intrinsic and basic human right to feed oneself. A right to grow some or all of your own food. Though my own hens are family pets, my family now depends on their eggs, which are a luscious addition to the vegetables and herbs grown nearly year-round in my kitchen garden. My partial self-sufficiency is a source of personal pride and an inspiration to my friends and family. My urban chickens are a symbol of my commitment to be more self-reliant. Though I still shop at the grocery store, I now try to feed my family more from the backyard than the supermarket. Don't crack up; I feel empowered by "growing" my own eggs. So do many fellow chicken keepers.

I've lived in big cities and busy suburbs all my life. If I'd known all along that keeping a few hens in my backyard was not only legal but also fun and inspiring, I would have had chickens long ago. Now that I do know, I'll never be henless again.

This book's instructions for starting and maintaining a flock are distilled from my experience in raising hens in the city. They will also hold true for those of you who live in the suburbs, in small towns, or out in the country. If you worry that you'll be the oddball chicken keeper on your block, rest assured that there are more chicken enthusiasts out there than you know. If articles that appeared over the last year in *House & Garden* and the *Wall Street Journal* are any indication of chickens' rising popularity in urban neighborhoods, then the Power of Fowl is being realized. Forget the phoenix — chickens are rising!

# From City Chick to Urban Chickens

HICKENS IN SAN FRANCISCO, Boston, and Seattle? Why not? Chickens in the suburbs and bedroom communities of Saint Louis, San Diego, and Chicago? Most certainly. Keeping enough hens to provide a family with fresh eggs every day is not a new idea. But it is a novel idea in this day and time, riding on the wake of a growing trend toward increased quality of life and greater control over the little things in life — like the eggs we eat.

At any given moment, there are 10 billion chickens clucking around the world. Hens worldwide lay close to 700 billion eggs each year. Most chickens live on large commercial farms. A few live on small farms or rural homesteads. Fewer still live in cities, suburbs, and towns. But this is changing, in my own neighborhood and in neighborhoods across America, as more and more people are keeping small flocks of chickens tucked away in their city and town yards and gardens.

My decision to keep chickens — hens, actually — for eggs and amusement in my tiny Portland, Oregon, yard began with childhood reminiscences that inspired a landscaping project that evolved into a personal philosophy of self-reliance and sustenance and more control over the source of my fresh foods. All because of chickens? Sounds crazy, I know. But I've never felt saner. Those ubiquitous yet taken-for-granted chickens have changed my life, and I love them for it.

That's right, I love my chickens! They are a part of my family just like any other loved and loving pet. Those of you who are not yet thrilled by the idea of chickens in your backyard deserve an explanation.

If ever there was an unlikely person to enjoy the company of chickens, it's me. My exposure to chickens as I was growing up was minimal. Ironic, considering that my father grew up the son of a butcher on a farm in Poland, and my mother was raised on a farm in France. Despite my pro-chicken gene pool, I grew up instead to be a Total City Chick.

I was born and raised in Los Angeles, where "chicken" was the derogatory name for boys who wouldn't fight in the schoolyard. In the "planned unit developments" where my family lived, I rarely saw live chickens, except at petting zoos and (for brief moments) at Chinatown meat markets. The only other time I saw chickens in any other state of being was at the dinner table, trussed, stuffed, and crusted golden brown. I never thought about chickens much unless deciding how I wanted mine cooked — fried, barbecued, sautéed, broiled, braised, stewed, or baked. Given chickens' relentless versatility as a main course, years later I shouldn't have been surprised that this selfless, utilitarian bird would be so adaptable to living in my modest urban backyard.

Now I live in Portland, the biggest city in Oregon. Here, "chicken" means *chicken*, the feathered bird of barnyard lore. My neighborhood is the dense, busy southeast quadrant known to locals as "the Southeast." Despite appearances — the houses tend to have broad front porches and long driveways — homes here are "cozy" (that is, small and very close to neighbors' windows). The yards are also cozy, having as many as five adjacent backyards. Gardeners need creativity and verdant screens to make these cramped spaces seem private and bountiful.

A stroll down most streets in Portland provides plenty of evidence that gardening is seriously embraced as a community-wide pastime. Front yards of the Southeast bungalows are lush, loved, and crammed full of broadleaf rhododendrons, bamboo, and flowering bulbs like daffodils, tulips, lilies, and dahlias. Folks make the most of their tiny city yards.

I was bit by the gardening bug when I bought my house. The 1908 Craftsman bungalow is big on old-fashioned charm and detail and *small* on yard space. Original features include an entry hall with leaded glass windows, doors and ceilings framed in wide, ornate moldings, and doorknobs made of glass, brass, and porcelain. The curb outside features half-dollar-sized iron rings to which inhabitants and visitors to this house tied their horses nearly a hundred years ago. In exchange for all the historic structural detailing, I had to give up yard space: My home has a very small lot, not wide enough for a driveway. Nonetheless, it was and is *my* land. There is nothing so grounding, so "American Dreamish," as buying a home and doing what you want with it. I felt a greater connection to the land and the structure than I'd ever felt as a lifetime renter, and

Poultry Tribune, circa 1940.

Poultry Tribune, circa 1940.

more home-oriented hobbies than I had imagined could exist began to interest me.

Shortly after settling in to my modest bungalow with the postage-stamp–size yard, I got to work. Like my neighbors, I've had to landscape efficiently. I've managed to cram in raised flower beds, a fern bog, a grass hill, a berry patch, and various trees, shrubs, bamboo, and bulbs. I installed two concrete patios (one covered), two winding stone paths, a birdbath, and a bucket pond with koi and water plants. Once I got started in the garden, there was no stopping. The more I planted, the more I had to tend to, and the more I enjoyed my garden. I had thought that once you plant a garden, you watch it grow from afar (boy, was *I* green!). I found out that gardening is an interactive experience, with the grower right there in the front trenches, the raised beds, and along-side the irrigation hoses.

When I started to garden, I grew only shrubs and flowers. A couple years passed. While lament-ing some leaf mold on a rosebush, I had this crazy idea. Perhaps in addition to my plants and flowers, I could grow vegetables. The thought of pressing a dry-looking seed into the ground, watching it sprout, grow, and mature, and eating whatever had grown was excit-ing to a City Chick. Mom had grown tomatoes, potatoes, and grapes in the garden I grew up in. Now, I thought, it was my turn.

Because the yard is small, I had to make room for a veggie patch. My spouse rented a jackhammer and chiseled away one-third of the uncovered patio. We recycled the concrete pieces into garden stepping stones, put up wood framing to create a raised bed, and hauled in what seemed like a hundred wheelbarrows' worth of premium soil for the bed. After a couple seasons, I had a small vegetable garden crammed full of fresh herbs, cabbage, eggplant, onions, garlic, tomatoes, squash, green beans, and several varieties of lettuce.

There I was, a City Chick growing and eating my own vegetables right there in the middle of a metropolis. And the best part was that growing fresh veggies, while not particularly difficult, was very rewarding. The more herbs and vegetables I grew, the more empowered I felt to grow others. The jackhammer came home again, and I soon had another patch of dirt in which to grow fresh food. I planted with the passionate fervor of Scarlet O'Hara. Harvests abounded. Tomatoes tumbled from tomato cages; lettuce lined up thick as a lawn; jalapeños jumped off the plants.

As I ate more and more fresh food from my garden, I noticed that I became more and more picky about how food tasted. After all, if I was taking the time to chew it, it had better taste good. I stumbled upon an important concept: *Fresh and organic = good.* Since the prices of certified organic vegetables are higher than the current worth of my retirement stock account, and probably will be for some time, growing my own vegetables is a valuable endeavor.

Without realizing it, I started hearing voices. Voices talking about chickens. My spouse had often told me childhood stories about keeping chickens while growing up in Oakland, California, always speaking fondly of those city hens and their antics. Also, my mother and father never stopped talking about the chickens of their youth. Though I was unaware of it, for years a very strong "chicken vibe" reverberated along my lifeline. In fact, my family was always talking about chickens. Chickens, chickens, chicken talk all the time. The only person not talking chicken was me.

**Chick Chat**

Chickens can become part of your family like any other loved and loving pet.

■ ■ ■ ■

The Main Rule for keeping urban chickens: No roosters allowed.

I never gave this much thought, until One Day.

One Day, while walking in my neighborhood, I was surprised to come upon some chickens. There they were, strutting and digging around in a narrow picket-fenced side yard. It was a serene setting, with irises and azaleas growing in the coop and the colorful chickens in the foreground. Walking on, I saw that their owner had built them a quaint henhouse inside an enclosed run. Metal sculptures were hung near and above the coop, giving it an artsy feel. Something about the sight of those chickens on a city street corner was really cool. And the birds (about four of them) were rather cute, walking about nonchalantly, then bending over with force and precision, beak picking up whatever beady chicken eyes saw crawling through the dirt.

## NEST BOX NEWS
■ ■ ■ ■ ■

Chickens raised in small city yards generally have more "personal space" than chickens raised on large commercial farms.

That evening, I was baking and thinking about those chickens. On account of my ravenous sweet tooth and my disappointment with most commercial cakes and sweets, I bake just about every day. As I finished cracking eight eggs into a bowl for a flourless chocolate cake, I had a rich thought: This cake would taste so better with fresh eggs! A neighbor had a small flock down the street. My spouse had had chickens in a dense bedroom community in California. And thanks to Mom and Dad's nostalgic reminiscences, I had been unknowingly indoctrinated in chicken-speak and knew quite a bit about those birds. Given all this, why couldn't I have my very own chickens here in Portland?

My next not-so-rhetorical question was "Is it legal to keep chickens in the city?" As I worked at that time as a paralegal, I knew the answer would be found in my city's municipal code. A municipal code or city code is a city's codified or statutory governing law. Depending on where you live — a city, town, or township, for example, the codes may also be known as *ordinances, codes of ordinances, regulations, revised municipal codes,* or *general provisions.* Many cities have their city codes posted on the Internet. Portland was no exception. After a few clicks online, I found my city codes and ordinances pertaining to chickens, which told me that within the city limits of Portland, residents can keep up to three hens without a permit. Additional hens could be kept, though a $25 annual permit was required. Also, the city codes had certain restrictions on keeping the coop area away from neighbors' kitchen windows. The final rule: No roosters allowed. That was okay with me, because I wanted hens for eggs, not for baby chicks.

When I told my spouse I wanted chickens, I expected the usual response to one of my wacky ideas: a slight, wary smile accompanied by that "Oh . . . *right"* look and a sad shaking of the head from side to side. Instead, my spouse grinned broadly, pumped a fist in the air, and shouted, "All right! Chickens!" (Actually, the reaction was somewhat more muted, but positive nonetheless.) After seventeen years, I guess we can still surprise each other.

I surveyed the backyard. This took about a minute. On a 3,300-square-foot lot, there's not a lot to survey. In nine years, most of the front and back yards were landscaped. The only thing left undone was a narrow, unused side yard that looked rather tawdry in contrast to the rest of the garden. It was choked with weeds, scrap wood, chipped bricks, and bad dirt. (I did say tawdry, right?) It was so messy and needed so much work that I had no choice but to ignore its existence.

Even if that unpromising patch of earth was cleaned up to make room for gardening, sunlight was restricted on that particular side of the house. There wasn't hope for me to grow much in this area. Before I had started "thinking chicken," the only appealing option was to completely cement over this side yard and build a shed to store the lawn mower and garden tools. Exciting, huh? Luckily, I had a healthy streak of procrastination running through me. I had done nothing with the side yard for almost ten years. After all, it was a nearly unworkable area. Little did I know that this meant it would be perfect for a coop and a small flock of city chickens.

Keeping in mind the setback requirements of our city code, we planned a relatively spacious, sturdy, covered coop and henhouse. With my spouse's mental blueprints and amateur knowledge of basic carpentry, the structure slowly rose out of the debris that was once a messy side yard. We worked in our spare time in the evenings and on weekends, and the project took about three weeks. We sunk the main support posts, put up the framing and roof, cut and fit four doors for the coop and henhouse, attached the requisite chicken wire, and painted the entire structure. It took a couple more days to clear out the debris behind the coop and to set up a dry place to store straw and chicken feed.

I had been studying my Murray McMurray Hatchery catalog for months. I could almost see the chicken illustrations in my sleep, and I had about a dozen favorites and alternates already in mind. After the coop and henhouse were up, I called a few local feed stores to see who had baby chicks. Chicks are seasonal in availability; thankfully, it was spring, when they're most available. We found a store with the breeds we wanted. We got the chicks. Over the summer, we watched them grow up into chickens. Ever since then, we have had fresh eggs almost every day. And my cakes and cookies have never tasted better!

At the time of this writing, I have three big lovely hens: Lucy (a Rhode Island Red), Zsa Zsa (a Barred Plymouth Rock), and Whoopee (an Australorp). I refer to the constituents of my own urban flock collectively as "the Girls." The Girls joined my family as chicks, I raised them through pullethood (pullets are adolescent hens) into their adult existence, and I will have them until they retire. ("Retirement" for chickens is not necessarily a euphemism for chicken dinner. See chapter 2 for details.)

So there it is, the tale of my transformation from city chick to urban chicken keeper. Each day my little flock of hens helps me find the taste of the good, simple life. And I feel a sense of pride and security in being able to feed my family and me from my urban garden, both vegetables *and* protein, thanks to my chickens.

# CHAPTER 2

# Why Keep Chickens?

UNTIL ABOUT 50 YEARS AGO, It was common to keep a few chickens on one's property, however modest the parcel. Cities had not yet spilled over onto adjacent farmlands and rural residences. Folks grew a good portion of their own food. Chickens were an integral part of the family food chain. Small flocks of chickens coexisted with "kitchen gardens," compact plots of earth growing enough greens and vegetables for a family. Three, four, perhaps a dozen hens provided the household with eggs and meat. Fresh food was available right outside the back door.

In those days, chickens were part of everyday life. With the rooster's crow at dawn, hens in the henhouse would stir from sleep, as would the humans in the peoplehouse. During the morning, somebody would let the hens out into the yard and collect any "early bird" eggs. The chickens roamed around all day, scratching for bugs and grit and laying eggs in secretive nooks of the yard. In the late afternoon, someone would toss a few handfuls of grain into the coop to

## Chick Chat

Two to four hens will provide plenty of eggs and amusement for a small household.

■ ■ ■ ■

The color of a hen's ear lobe roughly corresponds to the color of her eggs.

■ ■ ■ ■

Once your urban flock is established, daily chicken care is minimal.

lure back the free-ranging chickens, and the kids were sent out to search for the eggs.

Life got busier over the next several decades. Personal time and space, and especially backyard size for single-family homes, decreased. *Progress* became the defining buzzword and underlying foundation of a "good and successful" life. To pursue Progress, whether by choice or at the whim of destiny, people moved away from their spacious rural homes and into compact and convenient urban and suburban communities. With their migration, one of life's most prized possessions — personal self-sufficiency — was lost. As a consequence of shrinking yard space, garden flocks or family flocks of chickens disappeared from household yards and were culturally banished to the margins of farm life. At the same time that chickens flew the household coop, chicken ranching was growing into an agribusiness. In place of family hens roosting on a picket fences or scratching beneath the kitchen window, gigantic chicken farms and egg-production "factories" sprawled across the landscape. With Progress nipping at the heels of the mid-twentieth century, chickens became more than family food — chickens were big money for big new chicken businesses. Times had changed for chickens and for Americans. In France, chickens have historically been a symbol of good fortune and prosperity. In the United States, chickens have become a nostalgic icon of a way of life now gone by.

The past two or three generations of Americans — the Baby Boomers — never considered keeping a small flock of chickens as pets or private egg suppliers. Chickens were considered dirty, noisy, stupid, and needing more care than busy city folks in pursuit of Progress could offer. Progress, together with then-adolescent Convenience, convinced us that easy-to-obtain, commercially produced eggs were just as good as home-grown fresh eggs.

When I was growing up in the early 1960s, my parents and grandparents spoke fondly about the chickens they had while they were growing up. Mom always smiled when she recalled her chicken chores: feeding the hens and finding the eggs. The way Mom told it, having chickens sounded like a fun pastime, not a burden. After all, what kid doesn't like animals? Especially animals that created an egg treasure hunt every day. But Mom, Dad, and Grandma all grew up so long ago and so far away on small farms or in homes with oversized yards. I was a kid living in a crowded bedroom suburb of Los Angeles, plotted for maximum occupancy and minimum greenery and animal life. Nobody had to grow anything. Fluorescent-flickering supermarkets lined the wide, asphalt boulevards. Prepackaged produce, eggs, and meat were readily available, sometimes 24 hours a day. Nobody I knew had chickens. Chickens were synonymous with "country life," which was out there, somewhere, far away from where I lived in the infinite suburbs.

Consequently, our Progress-driven culture evolved to a chicken-deprived culture. Generations of Americans became completely removed from chickens and common chicken facts. Today most people don't even know that hens lay eggs *with or without* a rooster on the premises, and that brooding is not just something you do when you lose your cell phone. Folks have forgotten, or no longer believed, that chickens could be kept in

*Poultry Tribune*, circa 1940.

a modest yard to provide one's family with fresh eggs. What was once a simple part of life had taken on a complicated reputation. After all, don't chickens need a lot of space and care? Don't they make a really big mess? Don't they make a lot of noise?

The answer to all of the above is *no*. Chickens and the typical city or suburban garden complement each other if set up and maintained the right way. A few chickens are easy to care for and provide enthusiastic pest control and loads of free fertilizer. They also are guaranteed to provide you and your family with hours of relatively free entertainment. Depending, that is, on how you define "entertainment" — my family is easily amused.

Chickens don't need a lot of space for their home and hangout. Three standard laying hens need a henhouse the size of a large doghouse. They also need an enclosed outdoor coop (pen), at least four times the square footage of the henhouse, in which to stretch their legs. As for the time spent actually caring for chickens, it takes about 10 minutes each day to feed and water the hens and collect their eggs. Once you realize how little time is needed from your busy, overscheduled, Progress-driven day to keep chickens, you'll make the necessary time. Caring for your hens will subtly evolve into a pleasurable routine and brief respite from a hectic day.

That is exactly what is happening in towns, suburbs, and cities across America. For example, where I live in Portland, hens and henhouses are hatching all over in the modest yards of city homes. These urban chickens and their coops are most noticeable in my neighborhood, a cosmopolitan community with a hip, eclectic blend of businesses — cafés, beauty and bath shops, bead stores, bird suppliers, chiropractors, and art galleries, just for starters.

In the middle of this dense, busy neighborhood, there's a whole lot of clucking going on. Quiet streets tapering away from the main arteries carry the sound of a hen's happy cackles after laying an egg. A quiet flock of bantams (miniature breeds of chickens) scours a grassy parking strip for beetles and worms. Behind flowering pink dogwoods, a compact coop built on stilts has plump hens with vivid black and red plumage scratching underneath.

Portland residents and all other city, town, and country animal lovers with farm fresh tastes are bringing small flocks of chickens back into the garden. And for good reason, too. No other farm animal is as adaptable to small-space living as is the chicken. Miniature pigs are small in name only and in reality are quite heavy on the hoof. Cows are too large, goats are too destructive, and horses need special and spacious accommodations and recreation. But chickens can join a household with no more difficulty than a cock-atiel, rabbit, or reptile. Chickens don't get too big, don't need specialized care, and don't need to be taken out for a walk. How about that? Chickens are easy, chickens are fun, chickens are *in*.

Lately, no matter where I go, everyone is talking chicken. On the bus, riders discuss the different shades of egg colors laid by their hens. In the grocery store, a little girl runs up to her mother with a head of lettuce, pleading, "Can we get this for the chickens, Mom?" Local newspapers carry feature articles about chickens, their coops, their owners. Slowly, surely, chickens are migrating in small flocks back into hearts and gardens all across North America.

Surprisingly, most large cities in all 50 states of the United States permit residents to keep two, three, or more hens. (For a partial list of many big cities that allow residents to keep chickens, see the appendix.) Most municipal codes permit chickens in residential areas, subject to one or more codified stipulations (aka "the Law"). For example, the chicken house may be required to be a minimum distance from neighbors' residences. Nuisance codes prohibit the indefinite clucking of happy hens, and health regulations specify coop sanitation standards, such as cleaning the coop once or twice a week, and more often in summer. Most city codes aren't that restrictive. In fact, city codes that pertain to keeping chickens are quite reasonable. If you don't keep too many chickens, you *can* raise chickens in the city.

Beyond the realm of keeping of a few chickens for pets or eggs exists the world of *fancy poultry.* No, we're not talking about chickens in tuxedos and slinky gowns attending a black-tie charity event. Think *chicken show* — a grand, competitive get-together for folks who have elevated their chicken ownership from mere amusement and incidental practicality to a deep and abiding appreciation for breeding diverse, interesting, and beautiful chickens. These "professional" chicken lovers are known as *chicken fanciers* or *poultry fanciers.* They raise their purebred birds according to the American Poultry Association's *American Standard of Perfection,* the undisputed blueprint for the standard characteristics of each officially recognized breed. Each year, several hundred poultry shows are held in the United States and abroad where chickens and their owners strut their stuff.

After raising two or three chicks to adult hens, smiling at their antics, and collecting their eggs, you might become really excited about chickens. You may even want to raise poultry for show or for meat. My advice: Don't do it! Raising chickens to show or to butcher for the meat is an entirely different experience from keeping a few hens to provide enough laughs and eggs year-round for you and your family. Raising chickens for meat or show involves a significantly greater amount of time, expense, and space than keeping a small flock of pet chickens. There's also the rooster problem. The most economical way of raising chickens is to keep a rooster for fertilizing your hens' eggs. Because nearly all cities and towns have laws prohibiting keeping roosters within their limits, you most likely cannot have a rooster, and therefore no baby chicks. But that's okay — you will have plenty of fresh eggs to enjoy and share with family, friends, and neighbors.

However, if you are absolutely sure that you need more chickens, gratify your need, but not at your neighbors' expense. A few hens are charming; more than a dozen can get quite noisy, even without a rooster crowing all day. To keep you, your neighbors, and your chickens happy, keep larger flocks of chickens on oversized lots or on acreage as far away from neighbors' windows as you can. Never keep more chickens in your residential yard than is permitted by law. Folks who flout the law and the common sense restrictions about keeping large numbers of fowl on too small a parcel of property create a headache for neighbors and city employees and give city chickens a bad name.

One final note. While referring to *chickens* throughout this book, what I actually mean is *hens,* because roosters usually aren't allowed within city and town limits and outlying areas. So, when I mean *rooster,* I'll say *rooster;* otherwise, I mean *hen.*

## Chicken Care Is Easy

The number one misconception about keeping chickens is that chickens are difficult and time consuming to care for. I've had parakeets. I have chickens. The chickens are easier. They're also more practical (ever bake with parakeet eggs?) and physically heartier (chickens don't drop dead in slight drafts).

Like dogs, cats, and other household pets, chickens have basic needs: shelter, food, proper sanitation, and some exercise. But once a small flock of chickens is established in your yard, for the most part all you'll have to do is watch them and eat their eggs. The biggest investment of time for your chickens comes at the front end of the flock — designing and building chicken facilities.

A chicken's home is composed of the *coop* (a fenced outdoor area for chickens) and the *henhouse* (a small shelter located inside the coop). The size and style of the coop and henhouse depend on your creativity, your available yard space, the setback requirements of city code, and the number of hens you expect to keep. Chapter 5 offers advice and instructions for building a coop and henhouse.

The henhouse and coop should have clean, dry bedding material to absorb moisture and odor. I recommend spreading a layer of cedar chips and topping it off with a layer of chopped straw. The combination of cedar and straw absorbs moisture, cancels out some of that "fresh barnyard smell," and is easy to muck

**Chick Chat**

Chicken chores boil down to keeping the coop and henhouse clean, providing plenty of fresh food and water, and collecting eggs.

out. Depending on the weather, and no less than once a week, the coop and henhouse will need to be cleaned out. Cleaning the coop takes no more time than cleaning out a hamster cage or bathing your dog. See chapter 7 for details.

You'll have to check every day to make sure the chickens have plenty of food and water. The best way to ensure your hungry hens never want for food is to fill a large stainless-steel feeder several times a week with chicken feed. Of course, it's also fun to hang out with your chickens and hand-feed them table scraps (plain breads, pastas, vegetables, and certain fruits). Your hens' ravenous appreciation won't go unnoticed.

Although not known for being great athletes, chickens do like their daily exercise. For this reason, a roomy coop is important. Chickens should have plenty of space to dig, dust themselves, and flap their wings. But even if you have a spacious coop, you should let your chickens out into a fenced-in section of your yard a couple of times a week, even if only for a couple of hours before sundown. They will run, flap, hunt for bugs, and otherwise entertain you. Who needs television?

## Fresh Eggs

One of the reasons for keeping a small flock of chickens is the delicious eggs you will get. Fresh eggs are rich. Fresh eggs are fabulous. Fresh eggs *rule!* Once you've tasted one of your hens' eggs fresh from your own garden, store-bought eggs will forever after taste bland and light. Some people — especially those raised on a lifetime of commercial eggs — think this yolky decadence is too rich to eat. If you are one of these people, then you can give away the eggs and simply enjoy the chickens. But if you're like me, you'll become a fresh egg snob that gazes upon commercial eggs with patronizing tolerance at best, and with outright mockery on days when you are feeling a wee bit cranky.

Why am I wild about fresh eggs? Just crack one open and take a look. You'll notice first, in trying to break open the egg, that the shell is resilient and tough to crack. You can use a bit of strength when cracking a fresh egg; it won't collapse in your hand like those thin-skinned commercial eggs.

Crack the fresh egg into a heated pan bubbling with butter. Your eye will go right to the yolk. Oh, the yolk! If you're a yolk person, like me, you'll be amazed by the bulging round mound of yolk that stands tall in your fresh egg. You'll be entranced by the much darker and bolder color of the yolk — it's more orange than yellow. Then the egg white gets your attention. Rather than thinning out and waning away from that robust yolk, the egg white holds its own in the frying pan. It's more viscous than its commercial counterpart and spreads more slowly. As it cooks, it becomes the whitest egg white you've ever seen. Fork it in, and you'll see that it tastes as buttery as it looks.

The freshness of homegrown eggs fills out the flavor of your cooking. Cakes and omelets come out with more flavor and a denser, richer texture. Pancakes taste sweeter, and custards are silkier.

Chicken eggs are as chickens eat. Feed your chickens organic pellets and scratch, and you will have organic eggs. Nonorganic chicken-laying feed is also available — at $5 less per 50-pound (23 kg) bag, your hens will get all the nutrition they need, and your eggs will be fresh, though not technically organic. Toss in some kitchen scraps like leftover greens, vegetables, fruits, and a little day-old bread and pasta. Your chickens will be so happy that they will make their eggs taste even better for you!

A note about the term *organic:* There are relative degrees of organic, as specified by recent federal regulation. Basically, *organic* means something grown without the use of pesticides or other

*Poultry Tribune,* circa 1940.

synthetic chemical agents. You may consider your eggs to be organic because you are "growing" them in your backyard, even if you are feeding your hens the less expensive nonorganic feed. Organic purists would argue that organic eggs are truly so only if the hens are fed only certified organic feed, are fed no nonorganic scraps, eat no grass or weeds from a nonorganic lawn, and receive no antibiotic treatments. You, in turn, might want to pull out your hair. In truth, at this time in our history, the definition of *organic* is not only relative but also personal. The bottom line is that if you feed your chickens a healthy, balanced diet, the eggs they lay for you will always be sweet and tasty.

Hens lay an egg about every 25 hours. Sometimes they skip a day of egg laying. (They skip several days or weeks of egg laying when they molt in the fall.) It's usually nothing to worry about, unless accompanied by visible symptoms of discomfort or unusual behavior. Sometimes Lucy will skip two days of laying, then lay two eggs the following day. The other two hens will skip a day here and there. Altogether, the Girls provide me with up to 20 eggs per week. Counting the cost of their feed, that works out to about 50¢ per dozen of fresh brown eggs.

Most hens lay brownish eggs, sometimes with a pink, red, orange, or lavender hue. You can tell what color egg a hen will lay by the color of her ears (yes, chickens have ears). They are located behind and slightly below a chicken' eyes and can be partially feathered over. If you gently push the feathers aside, you will see the skin-like texture of the ear opening. This color roughly corresponds to the color of the eggs the hen lays. Hens with reddish or pink ears will lay brown eggs; pale-eared chickens lay white eggs.

Zsa Zsa, my Barred Plymouth Rock hen, lays slim, dark pink eggs with a brownish tinge and a flat finish. The eggs of other Rocks I've seen ranged in color from light to very dark pink; some had a shiny, not flat, finish. My Rhode Island

**BIRD WORD**

For truly organic eggs at just about 50¢ per dozen, feed your hens organic laying feed and organic greens.

hen, Lucy, lays reddish brown eggs with tiny brown speckles. Whoopee, the Australorp, lays glossy eggs that are brown with dark red and orange hues. After collecting several days' worth of eggs, I put the bounty in a white ceramic bowl on my dining room table, where the mound of fresh eggs glows like a multifaceted brown-hued jewel.

Check the henhouse daily for eggs and, if you can, collect them promptly after they are laid. They need to be brought in and refrigerated within the day to prolong their freshness. See chapter 7 for more information on egg handling.

## NEST BOX NEWS

■ ■ ■ ■ ■

Hens lay an egg about every 25 hours.

■ ■ ■ ■

Collect eggs every day, and store them in the refrigerator to prolong their freshness.

Having fresh eggs is a joy and a privilege. A joy because nothing else tastes like a fresh egg, and a privilege because not everyone enjoys the company of hens like my friends and I do. As more folks discover how easy and rewarding it is to keep chickens and it again becomes common to keep a garden flock, I hope for more egg joy and less egg privilege for everyone.

In deference to folks who may not like eggs or who are allergic to them, forgive the egg-centric tenor of the preceding paragraphs. For you folks, bird beauty, not egg production rates, is reason enough to have an urban flock. You can appreciate your chickens as pets and mobile lawn décor and also give the gift of eggs to family and friends.

## Pest Control

Chickens love bug hunting. More to the point, they *live* for bug hunting — it's a hen's Holy Grail. Something about digging up a plump worm or juicy beetle drives the Girls crazy. Which is fine by me, as I am not particularly fond of worms, beetles, or any other wriggly or creepy life-forms that could potentially crawl up my pant legs or into my ears.

Since I started keeping chickens, the pest population in my garden has noticeably decreased. Creepy crawlies cower in fear before the Girls. Earwigs, centipedes, and beetles shake on all their legs as my hens thunder up to the rock or dirt hill concealing the unfortunate insects. A flip of the beak, and bugs be gone. More protein for the Girls, fewer bugs for me to squash. A perfect match, and I can't think of a more natural approach to pest control.

However, when the Girls are out controlling pests in the garden, I have to control the Girls. In the excitement of the hunt, my bell-bottomed ladies can trample a garden plot of lettuce in three minutes flat. You wouldn't think that a bird or two could be so hard on a garden, but the Girls are no canaries! When you let your chickens range free in your yard, you may want to fence off — permanently or temporarily — anything you don't want crushed, dug up, or eaten. Or you can plant a garden resilient to chickens, such as a hardy, drought-resistant lawn bordered by evergreen arborvitaes, spiny conifers, and some rocks. (I'm exaggerating, but you get the idea.)

About pesticides — I no longer use them. I don't need to, because the Girls get the bugs. And because the Girls eat the bugs, I don't use pesticides; the chemicals might harm them. Chickens pick at and taste everything. So if you use pesticides or lawn or flower fertilizer in the garden, keep your hens in the coop for a few days afterward until the chemical additives have been absorbed or thoroughly washed away.

## Top-Notch Fertilizer

Whoever thought I'd be singing the praises of chicken poop? I am, and I'm not the only one. Chickens are walking nitrogen-rich manure bins. The Girls' manure is the envy of my gardening friends. When I fluff up the Girls' coop, I take the generous deposits of chicken guano and straw from beneath their sleeping perch and put it right in the compost bin. I store my leftover guano straw in extra trash cans that my friends line up for and haul away for their own compost bins. Nitrogen is an essential ingredient in great compost, and chickens are just full of it!

In the winter, I throw the guano-laced straw right on top of the fallow vegetable gardens. Then I let the Girls out into the yard, and they beeline right for the vegetable patches, where they spend glorious hours searching for bugs in the soft soil and digging the compost materials deep into the dirt.

Some chicken keepers put their flock in a small mobile pen to create a movable compost bin. They move the pen in the yard every few weeks, bringing it to areas where the soil needs to be worked and amended. The chickens till their own nitrogen-rich guano into the soil of the pen, fertilizing the land directly under their living quarters. These types of pens are best suited to larger yards that can accommodate a roving 3' x 4' x 2' (91 x 122 x 61 cm) structure.

Chickens are natural partners in a complete home recycling program. Since I've had chickens, no leftover fruits, vegetables, or breads go to waste. Instead, the Girls get it all, which delights them from the tops of their rough combs to the tips of their sharp toes.

## Chick Chat

Your chickens will gladly take over pest patrol in your garden.

■ ■ ■ ■

Chicken manure is a potent fertilizer.

Besides, all those lovely, fibrous scraps just make them poop more, which gives me more guano to go in the compost bin. Feeding kitchen scraps and leftovers to a flock of city chickens is a win-win event.

# Chickens in the Garden

Chickens are beautiful. Chickens are fun. Chickens are also a bit rambunctious. Left to run unattended in your garden, gentle hens take on the demeanor of roadhouse thugs. They break blossoms. They crush tender shoots. They pull up baby lettuce and lay siege to unsuspecting squash seedlings. They don't mean to; they're just a band of happy, clumsy hens.

When I first started keeping chickens, I'd let them out of their coop and into my garden. I'd watch them a bit, then go back into the house. Two hours later, I'd go back outside and wonder where my garden went. Left behind was a landscape of freshly dug and scattered soil, several shallow holes, and rootballs and rhizomes laid bare, vulnerable and drying up on the sunny topsoil. While I appreciated how thoroughly the chickens had aerated my garden, I was disappointed not to have any plant life remaining that could have benefited from their efforts.

I quickly realized that my hens became hoodlums when left unpenned and unattended. Still, I wanted to enjoy watching them browse through my garden on occasion without sacrificing years of nurtured foliage. I wanted them to keep eating all those delicious bugs, too, so I came up with a couple of ideas.

The first part of the plan was to limit the time the Girls had to themselves in the garden. Instead of letting them out for hours at a time, I let them romp freely for about a half hour before dusk. With a limited amount of time, the Girls can crush only so much in the garden. I don't take any chances with delicate plants like bleeding heart and maidenhair ferns and delicious greens like baby lettuce. I have a large yellow push broom and a wide orange rake that, happily for me, seem to strike fear in the hearts of the Girls, even when the tools are simply propped against a fence. I place the broom and rake across beds that are off limits. The hens, certain that these tools are instruments of chicken torture, tend to steer clear.

**BIRD WORD**

Make sure delicate garden plants are protected from free-ranging chickens.

■ ■ ■ ■

To minimize their trampling, let hens out into your yard for only an hour before dark.

As a second part of the plan, I let the Girls out only at dusk. Part of the reason they dig holes so frenetically is to cool themselves; the earth below the sun-warmed topsoil is cool and soothing. At sundown, they aren't as likely to dig holes for their cooling comfort, because the warmest part of the day is already gone.

I allow the Girls long, leisurely hours in the yard when I'm gardening. This way, if I catch them digging near ferns or flowers unguarded by Bad Broom and Rake of Death, I can promptly shoo them away. Anyway, hanging out in the garden with my chickens is fun for me and for them. The Girls like to think they are helping me if they are directly underfoot. If I'm turning soil in spring, they crowd around like kids at an ice cream counter, waiting for me to turn up fresh, juicy worms. When I prune or trim plants and shrubs, the

*Poultry Tribune,* circa 1940.

Girls like to stand by the pile and taste the tossed clippings. They'll sneak up behind me, grab a twig from the debris, and run off with it as if hoarding valuable treasure.

More recently, my spouse came up with a great idea for allowing the Girls out in the yard without close supervision: a temporary fence. Think "baby gate" for chickens. It is made of heavy-duty netting that is 3 feet (90 cm) high and spooled like a bolt of fabric. One end is secured to a post near the fence in my yard, and the other end is stapled to a sturdy wood dowel that extends about 8 inches (20 cm) below the bottom of the net. I've sunk a PVC pipe in the ground at a point across the yard from the fence post. Before letting the Girls out of their coop, I roll the net across the yard and drop the end of the dowel into the open mouth of the PVC pipe. This divides my garden into two sections — mine and the hens'. The system is somewhat primitive, but it effectively seals off half of the yard from my cheerfully marauding chickens. You could also construct fencing with wooden dowels at either end and sink PVC pipes in several strategic locations so that you could fence off different portions of the yard.

The Girls maintain total rule on the narrower front portion of the temporarily fenced area just outside their coop. The rest of the yard belongs to me. In the front section, I coincidentally had plants

**This temporary, lightweight, portable chicken fence can be made for a minimal cost with items purchased at a local hardware store.**

that were resilient to heavy hen feet: established rhododendron, azalea, arborvitae, camellia, raspberry, mint, and other evergreen, woody shrubs (miniature conifers look adorable with chickens mingled among them). None of these plants has ever sustained much wear and tear during the chickens' pleasant pillaging. The larger back section behind the net fencing has all my delicate plants, including the ferns, peonies, and vegetables.

## Art, Pets & Entertainment

Yet another benefit of having a garden flock of chickens is being able to enjoy their beauty. That's right, their *beauty*. Considering the great variety of plumage colors, patterns, and styles, I think chickens are the most beautiful birds in the fowl world.

Unfortunately, the beauty of chickens has been eclipsed by their unglamorous and historically utilitarian role as meat and egg suppliers. A chicken's many colors and pleasing symmetry are generally overlooked by a chicken-deprived public. Keeping small flocks of colorful hens in the garden is like celebrating a live art form, with egg dividends daily.

Over the years, I've seen and read about many different breeds of chickens. While I really do love getting fresh eggs, I didn't pick the Girls solely for their egg-laying reputations. I picked them for their looks. I like them *big*. The bigger, the better. Being of petite stature, I seem naturally drawn to big things: big cars, big dinners, big rooms, big coffee mugs, and, of course, big chickens. Not that I don't appreciate the smaller bantam breeds. There's nothing quite as elegant as a snow white Silkie or a regal, fancy-tailed Japanese, two of the few true bantam breeds. But for me, the sight of an 8-pound Australorp hen heaving her broad breasts toward me during her rendition of "running" is both impressive and amusing. See chapter 6 for more detailed information on selecting the right breed of chicken for your urban coop.

I also love the Girls' beautiful plumage: a dark red-orange Rhode Island Red; a jet-black Australorp whose feathers shimmer with green hues in the sunlight; and a black-and-white, herring-bone-patterned Barred Plymouth Rock. Their bright colors are vivid against my thick, rarely mowed lawn and contrast nicely with the paint on their henhouse — a bright yellow and blue, with royal purple trim. When the Girls are walking slowly, as they usually do, their pleasantly plump bodies float across the lawn like big, colorful koi in a pond.

As for stress reduction and relaxation, backyard koi ponds and bubbling fountains have nothing on a flock of urban chickens. After a stressful day at the office, nothing makes me feel better than heading out into the backyard, sitting down in a lawn chair on the patio, and watching the Girls stretch their scaly gams. Gazing blankly at my pets peacefully clucking around on the lawn takes my stress down several notches and always gets me laughing.

The Girls love when I talk to them; during their babyhood, I spoke to them often so they'd become comfortable around me. The tiny Girls would fall asleep in my cupped hands while I spoke quietly to them. Listening to hours of my chick-happy monologues while growing up accustomed the Girls to my voice. As full-grown hens, they still enjoy my voice — sometimes too much. One evening I let the Girls out behind their coop, an area beneath my dining room window. I opened the window and started talking to them, and they happily cooed and scratched below. Except for Zsa Zsa. She suddenly

## Chick Chat

Fresh egg yolks are dark yellow or bright orange because backyard chickens (unlike their commercial cousins) eat lots of greens and vegetables that contain beta-carotene.

jumped up at the window. She hovered briefly near the sill, flapping madly, before clumsily fluttering back down to the ground, grounded again. Perhaps she missed me and wanted to catch a glimpse. Or (more likely) she thought I was holding a fresh cob of corn and wanted to beat the other Girls to it.

While chickens are not wired for affection and loyalty in the same manner as a cat or dog, they are friendly and loving in their own way. They don't cuddle with me on the sofa in front of the television, but the Girls do show me chicken-style love. If nothing else, I am their favorite walking food dispenser. When the Girls hear my approach — chickens have great hearing and sight, but not great senses of taste and smell — they all run to the coop door. Their clucking picks up pace as they jostle each other out of the way for a better view of me. Well, I like to think it's me; they probably just have their eyes set on the lawn or any food goodies in my hand. When I open the coop door, they spill out like ecstatic concertgoers, each barreling her way to the front of the line on their rush to the lawn. If I'm in the coop and bending down to tend to chicken stuff, Zsa Zsa will jump up onto my shoulder and pull at my hair. Or she'll peck at my jeans or untie my shoelaces when I stand still for a bit. Lucy and Whoopee aren't as cuddly, but they do like to hang out close to me, even when they have the entire yard available to them. Again, I won't flatter myself — they probably just don't want to miss any potential edible handouts — but I enjoy their company just the same.

My friends, my spouse, and I have learned that watching chickens in the yard will bring a smile to anyone's face. What's so darn funny about them? Here are a few brief tales of the Girls' escapades. However, I recommend that you watch and laugh at your own chickens. Chicken humor, like most comedy, is a highly subjective experience.

CHICKEN TALE NO. 1:

## Chickens in Motion

The Girls come tumbling out of their coop and onto the lawn. Zsa Zsa suddenly turns around, then dashes forward. She stops on a dime and, without any encouragement from the other Girls, shoots a couple of feet straight up into the air. Now Lucy and Whoopee get revved up, probably as a result of Zsa Zsa's floor show. Soon three crazy chickens are running around and jumping up into the air for no apparent reason. Suddenly, perhaps embarrassed by their awkward and poorly choreographed modern dance outburst, the Girls try to trick me into thinking I've just hallucinated a flock of bouncing chickens. They freeze in their chicken tracks, glance nonchalantly from side to side, and then bend down to nibble on blades of grass, as if nothing happened. I watch them for some time, but the Girls continue to cluck and munch quietly without any further jumping for apparent chicken joy.

CHICKEN TALE NO. 2:

## The Henhouse Sideshow

The Girls gather under a butterfly bush and quietly, contentedly peruse for insects in the soil. All is quiet. Then a shrill, panicked scream pierces the air. I look up to see Zsa Zsa running off with a beetle kicking several pairs of legs wildly in her beak. The other Girls are in hot pursuit and are also screaming. Chicken screams sound like an elongated, screeching "cheeeeeeP, cheeeeeeP" done over and over again until it resembles a single, continuous ear-piercing shriek. Suddenly, Zsa Zsa's concentration falters in the cacophony, and Lucy deftly moves in. With a single, undercutting motion Lucy snags the beetle from Zsa Zsa's beak with her own. Lucy is now running across the lawn, screaming. Whoopee jumps out of the bushes into Lucy's meandering path, the beetle drops, and Zsa Zsa comes out of the backfield, picks it up, and runs off into the direct path of Whoopee. This went on for a while until the beetle escaped (not likely) or was eaten (snap, crackle, crunch!).

CHICKEN TALE NO. 3:

## Three Heads
## Are Better Than One

Several friends and I are seated around a picnic table on the lawn. The Girls are pecking the grass near our feet. Zsa Zsa jumps up onto the picnic bench next to one of my friends. Lucy jumps up next to me. Then Zsa Zsa jumps up onto the table. Lucy follows. The two Girls are strutting atop the table like animated centerpieces. They cajole for crackers. They beg for cheese puffs. They whine for beer. They are shameless, as chickens have no idea what shame is. My friends comply with my request never to feed chickens at the table. Zsa Zsa gets impatient and jumps up on my shoulder. Lucy mimics Zsa Zsa and jumps up onto my other shoulder. Everyone is laughing. Everyone but me. I'm held motionless under 15 pounds of shoulder-mounted urban chickens.

The Girls get more than their share of visitors. Company loves the Girls, and the Girls love all the company. Since chickens joined our family, friends and neighbors have been regularly inviting themselves over to visit. They say they are coming to see me, but I know better. After a few perfunctory greetings and hospitable words, the conversation always turns to the hens. "So . . . how are the Girls?" Pretty soon, everybody is marching outside to the coop to visit them.

Our community coop culture became evident even as the first pine studs went up to frame the coop. Friends and family took an immediate interest in the project. My neighbors' children came over, asked questions about chickens, and even participated in building the coop. Everybody was always asking, "Is the coop done?" and "Have you got the chickens yet?" It was chicken talk, all the time. My

Mom and Dad, themselves former farm folks, gave me loads of advice and then surprised me with pop quizzes on my chicken IQ. These sweet, simple-minded birds were the hot topic in all conversations with everyone we knew. My pet flock had inadvertently become a family and community project.

You may have a Ping-Pong table, a satellite television, or an RV. No matter how many extracurricular amusements you have, once you add a pet flock of chickens to your household, you'll never have another boring day at home.

## Pondering Urban Flocks

Henry David Thoreau, one of the nineteenth century's great American writers, original thinkers, and vanguard nature buffs, believed in self-reliance and simplicity. Throughout *Walden,* one of his most famous works, Thoreau links the striving to simplify life with actively participating in the act of feeding oneself. By simplifying the means and ends of sustenance and, therefore, self-reliance, Thoreau believed, we would be closer to nature, to Earth, and eventually to ourselves.

Quality of life meant everything to Thoreau. According to him, such quality could be achieved by trimming away the extraneous, complex, and distracting elements of daily life. He emphasized simplicity and respect for the natural world and its animal inhabitants. In his literature, he seemed in awe of animals, even a common sparrow: "I once had a sparrow alight upon my shoulder for a moment . . . and I felt I was more distinguished by that circumstance that I should have been by any epaulet I could have worn."

If Thoreau was so moved by a sparrow, imagine how he would have felt about chickens! Of course, chickens don't "alight" upon

*Poultry Tribune,* circa 1940.

anything, especially on one's shoulder, but they are one of the natural world's greatest little creatures. What domesticated animal can feed large amounts of people while not encroaching on large tracts of land in the process? What so-called food animal has such varied beauty throughout its breeds? What animal can participate in and enhance our daily routine while requiring so little in return? The chicken.

I understood *Walden* intellectually, but it wasn't until I started keeping chickens that I really understood what Thoreau was writing about. What happens when I run out of eggs? Do I get dressed, grab my purse and car keys, gun the cold car engine with a gush of fossil fuel, and rush down residential streets to the Super Mart for a box of uniform, fragile white eggs laid who knows how long ago? I don't think so! Still in my pajamas, I open the back door, walk the short distance to my backyard henhouse, and gather the fresh eggs that are waiting for me, thanks to my small flock of hens. This is a lot simpler than what I used to do, and much more enjoyable.

The reemergence of chickens in America's backyards is a testament to the longevity of Thoreau's values. In an age when we're more and more estranged from nature and simplicity, I'm confident Thoreau would approve of chickens in the city.

I had initially approached my egg-laying garden chickens as a hobby — a fun and practical hobby. Now, I can't imagine life without the Girls clucking and scratching around in the yard. In a simple and affirming way, my chickens are daily reminders of the good things in my life.

# Chicken Basics

**B**EFORE DESIGNING YOUR COOP and deciding what kinds of hens to get for your pet urban flock, you will need to know some basic "chicken speak." If you are anything like me — a suburb-raised, city-loving person — then your general knowledge of chickens is low. Woefully low. It's nothing to be embarrassed about. After all, over the past 50 years, chickens have receded from common sight in one's neighborhood, being relegated to the margins of rural life and rare enclaves of animal husbandry. With them went rudimentary chicken facts.

When people find out I have a small flock of chickens, they always ask the same question: "Don't you need to keep a rooster in order to get eggs?" Though it's tempting to say something sassy in response, I refrain, because once there was a time when I didn't know the answer (although I was very young at that time). The answer is no. The only thing you need for eggs is a hen. A rooster is necessary only if you want those eggs to be fertile, producing baby

chicks. Besides, city law usually prohibits the keeping of roosters in residential flocks of the cities and suburbs.

What about those red, serrated "hats" on a chicken's head — what are those called? Do chickens have shins? Is there really such a thing as pecking order? Before answering these and other frequently asked questions, I will address the oldest and most frequent chicken question of all time.

## Chicken History

Which came first, the chicken or the egg? While the answer to this question is incessantly debated, the origins of the domestic chicken are more definite, though not entirely precise. Scientists generally agree that today's domestic chicken *(Gallus domesticus)* had its origins in Southeast Asia between six thousand and eight thousand years ago. At that time in what today would be the countries of Thailand and Vietnam, four types of wild jungle fowl *(Gallus gallus)* existed. Some scientists (including, in his day, Charles Darwin) believe that the chicken descended solely from the Red Jungle Fowl. Other members of the scientific community posit that more than one of the four wild jungle fowl varieties may have contributed to modern chicken DNA.

Most scientists and chicken researchers agree that chickens were domesticated in India sometime between 4000 and 3000 B.C. These chickens weren't "chicken" but fearsome flocks of fowl raised by royalty for cockfighting. Because of their aggressive, single-minded focus on combat in the ring, roosters became a symbol of war and courage. Throughout ancient history, battle garments and clan insignia bore the image of a rooster with his head held high and chest puffed out, staring into the distance, unblinkingly awaiting whatever combat should come his way.

Chickens begin to appear in historical and literary references in China, Egypt, and Babylon between 1500 and 600 B.C. Even the Greek playwright Aristophanes referred to chickens in his dramatic works. While Alexander the Great is known for bringing chickens to Europe, the birds may in fact have arrived earlier, accompanying far-traveling trade merchants and wandering soldiers as their primary source of protein on the roads of battle and commerce. It is at this time that chickens forged their role as the first real "to go" meal in recorded history.

The ancient, vanguard chickens gradually evolved into contemporary classes of chickens. From Europe, chickens most likely emigrated to the Americas care of Christopher Columbus in the North and the Spaniards in the South. Once here, chickens became an important source of meat for the inhabitants of and immigrants to the New World.

Poultry popularity came to a head in what is sometimes referred to as the "Golden Age of Pure Breed Poultry." During the mid- to late 1800s, dozens of poultry clubs sprang up in England and the United States. These clubs guided and influenced the creation of then-new breeds of chickens that remain popular today.

# Definitions

Now that you have some hen history under your hat, it's time to enhance your cosmopolitan vocabulary with a few choice words about the bird.

## GENDER & AGE

**Chick.** A baby chicken.

**Chicken.** A type of domesticated fowl raised and kept for meat, eggs, and ornamentation.

**Cock.** An adult male chicken at least one year old. More commonly called a *rooster*. Keeping roosters within city limits is usually not permitted.

**Cockerel.** A juvenile male chicken less than one year old.

**Flock.** A group of three or more chickens. Most municipal codes permit at least three chickens to be kept per residence.

**Hen.** An adult female chicken at least one year old.

**Pullet.** A juvenile female chicken less than one year old. A chick is considered a pullet when most of its feathers have come in and it's about the size of a husky park pigeon.

**Sexing.** The act and the art of determining chick gender. The sexing of baby chicks is 90 to 95 percent accurate. Sexing chickens is a highly specialized vocation whose practitioners are steadily diminishing.

## PHYSICAL CHARACTERISTICS

**Bantam.** A small variety of chickens weighing 2 to 4 pounds (0.9–1.8 kg). Most bantams are miniatures of standard breeds, although there are a few *true bantam* breeds (e.g., the Silkie and the Japanese bantams), meaning there is no standard breed equivalent. Also referred to as *banties*.

**Comb.** The reddish, skinlike "hat" or crown atop a chicken's head.

**Dual purpose.** Chickens raised for both meat and egg production. Dual-purpose breeds tend to have calmer dispositions than strictly egg-laying or meat producing chickens do. They lay brown eggs, except for the English Dorking, which lays white eggs.

**Oviduct.** The reproductive channel down which eggs descend to the vent.

**Shank.** A chicken's leg between the thigh and the foot; comparable to the human shin.

**Standard.** Larger chickens that are medium to heavy weight. Chickens of standard breeds weigh anywhere from 5 to 14 pounds (2.3–6.4 kg).

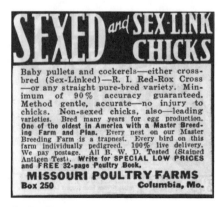

*Poultry Tribune*, circa 1940.

**Vent.** The chicken's bottom end, from which eggs and waste descend.

**Wattles.** Those red, fleshy, dangling muttonchops on either side of a chicken's beak.

## HOUSING & HEALTH

**Chicken run.** An outdoor area within the coop that provides a protected place for chickens to freely wander about; the equivalent of a dog run.

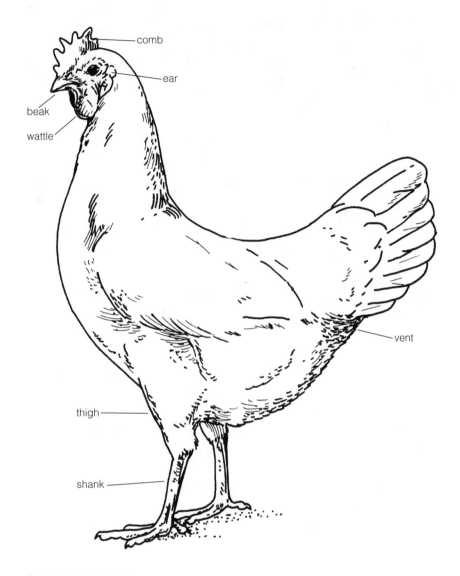

**Basic chicken anatomy**

**Coop.** An enclosed area for chicken habitation that contains a chicken run and a henhouse.

**Feed.** Chicken chow. There are different formulated feeds for chicks, pullets, and hens. Chick food can be medicated to prevent possible disease. Pullet food helps chickens grow and put on weight. Hen feed, also called *laying feed,* contains calcium for strong eggshells. Never feed hen food to chicks — the jolt of calcium in

*Poultry Tribune*, circa 1940.

younger fowl can damage their kidneys. Start giving laying feed to pullets when they are around four to five months old, right before they begin to lay eggs.

**Henhouse.** A shed or other small structure located within the coop where chickens sleep (roost) and lay eggs; the equivalent of a doghouse.

**Nest box.** A cozy, private cubicle in the henhouse where hens lay their eggs.

**Perch.** A small ledge affixed to an interior wall in the henhouse that hens roost (sleep) on. Hens like to sleep about 2 to 3 feet (60–90 cm) off the ground.

**Scratch.** The chicken equivalent of a snack. It's a mix of grains and cracked corn. Give it to your birds in addition to, and not instead of, fortified, enriched hen or pullet feed.

## CHICKEN BEHAVIOR

**Brood.** A hen sitting on her eggs to try to hatch them.

**Dirt bath.** Chickens' instinctive act of cleansing away and killing mites or parasites by digging a shallow hole, lying in it, and kicking up dirt onto their entire bodies. After a dirt bath, some chickens lie motionless and appear dead — they are relaxing after their satisfying "spa" treatment.

**Lay.** To produce an egg.

**Pecking order.** Social ranking of hens established naturally within the flock.

**Preen.** To run the beak through the feathers to clean and arrange them.

**Roost.** A chicken sitting on a perch to sleep at night.

# Chicken FAQs

Over the years, I've been asked all kinds of questions about keeping chickens in the backyard. While some questions are . . . well . . . unique, to put it mildly — Question: Can I walk my chicken on a leash in the park? Answer: Only if you want to attract a pack of wild dogs — most folks seek more ordinary information. The following are some of the most commonly asked questions about chickens and keeping them in the city.

## Do I need a rooster for my chicken to lay eggs?

No. Pullets and hens lay eggs without a rooster. A hen's natural plan is to release several hundred eggs during its lifetime. You need a rooster only if you want the eggs fertilized for the purpose of having chicks. Nearly all cities have ordinances against keeping roosters, because they are noisy. Contrary to popular belief, roosters don't crow just at sunrise — they crow all day long, frequently and loudly.

## How long do chickens live?

Depending on several factors — including your chickens' health, diet, and heredity — the average life

## NEST BOX NEWS

■ ■ ■ ■ ■

Chickens live about eight years, though some keep clucking for twice that long.

■ ■ ■ ■

Hens start laying at about six months. They lay the most eggs during their first two years.

■ ■ ■ ■

span of a happy hen is eight to ten years. Chickens have been known to live up to twenty years, though this is an exception.

## When do chickens start to lay eggs?

When they are about five months old. I've had hens start laying as early as four months and as late as six months of age.

## How long will a chicken lay eggs?

A hen's first two years are her most productive; she will lay nearly every day. Thereafter, egg laying slows down each year. We've had hens that laid eggs until they were 12 years old, albeit one per week. And I've read of a chicken (a Rhode Island Red) that laid into her seventeenth year! Most hens will keep on laying until their old age and demise, although they won't produce as many eggs as in their youth.

## What do you do with a chicken that no longer lays?

Old hens just don't lay as much as young hens. That's okay, because your hens are family pets, not objectified egg-laying machines. Your hen's gradually (and naturally) diminishing egg-laying capabilities are no reason to expel her from her flock and your home. As with all pets, you have a responsibility to keep and nurture your hens until their natural demise. If you don't think you can commit to your chickens through times of many and scarcely laid eggs, then don't get a flock.

When your senior chicken does finally expire, you can get another chicken or chick. Raise the chick separately from any remaining adult chickens in the original flock. Gradually, and with supervision, introduce pullets into the adult flock. Chickens are social birds and need to warm up to new flock members.

In the event that your love for your chickens was superficially related to the amount of eggs they laid and you don't want to keep your nonlaying hens, you have a few choices.

Your first choice is to eat your chicken. It's not a choice that everyone would make, but if you don't mind, then why not? This way, you don't subject others to your unwanted chickens. Although I don't think I could eat a chicken that I've named and that has been following me around the yard for years, chicken farmers and others

*Poultry Tribune*, circa 1940.

who raise large numbers of (unnamed) hens butcher and stew their chickens once the best of their egg-laying days are over. Stewing is the only way to cook an old chicken. Chickens over the age of one or two years don't provide the tenderest meat. Chickens raised for meat, like broilers — yes, they really are called that — are killed for meat when they are only six to eight weeks old. There really is something to the saying about "tough old birds."

If you don't want to keep your senior chickens but don't want to eat them either, put an ad in the paper offering them to a personal party or a petting zoo. There is always a chicken lover out there who may be able to squeeze another beak into a henhouse or flock.

*Do not ever* dump your chickens in the woods, in the park, or at veterinarians' offices. Not only is this irresponsible, but it also is cruel to the helpless birds. Your chickens, like all household pets, deserve better.

## What do chickens eat?

Chickens eat chicken feed, which is available at feed stores. Chicks and pullets that are not yet laying eggs should be given nonlaying feed; hens that are laying should be given laying feed. The difference between the two is that laying feed contains calcium, which is necessary for strong eggshells. Nonlaying feed does not contain calcium, because it can harm the kidneys of immature chickens. When pullets are ready to begin laying eggs, they graduate from nonlaying to laying feed. Chickens also eat *grit*, gravel or small stones that aid in the digestion of food.

In addition to the formulated chicken feed and grit, chickens love to eat fresh fruits and vegetables, like apples, melon, grapes, lettuce, tomatoes, corn, and spinach, and breads and pastas. My hens are crazy about cottage cheese; a spoonful (low salt or unsalted) lures them into the coop every time! Not only has our household food waste diminished to zero, but all the good food that goes into the chickens makes the eggs sweeter.

While you can give your hens vegetable leftovers and discards (like the outside leaves of a lettuce head), do not ever give them spoiled food. Also avoid giving your chickens any food that might add strong odors or flavors to their eggs, like cabbage or garlic.

For a complete roundup of chicken nutrition, see chapter 7.

## Do chickens have teeth?

No, they don't. Their beaks don't do the chewing — their gizzards do. When a chicken eats, food goes into its *crop*, a pouch in the throat that holds food. The crop delivers food to the *proventriculus,* or true stomach, where hydrochloric acid and digestive enzymes are added. Finally the food passes to the *ventriculus,* or gizzard. The gizzard is a muscular pouch that contains grit. Grit acts like a masticating agent, mashing up the ingested food.

## Do I have to bathe my chicken?

Not unless you really, really want to — and I can tell you from personal experience that the chicken *will not want to take a bath.*

Chickens are hard enough to hold still under dry conditions. Try holding them still during a warm bath and blow dry!

Mother Nature takes pretty good care of hen hygiene. Chickens keep clean through preening and dirt baths. A dirt bath removes any mites from a chicken's feathers. Preening removes fine particles of dirt and dust. However, chickens participating in a poultry show are bathed and groomed so that their natural good looks shine even more in the ring. Unless you are showing poultry or your chickens have shown an affinity for water, I'd refrain from bathing backyard hens. The chickens in your urban flock aren't going anywhere, so they don't need to get gussied up. Plus, a wet chicken is susceptible to colds.

## Can chickens catch colds?

Yes, they can. If you see that your chicken has the sniffles or is sneezing (yes, chickens sneeze), mash up a clove of fresh garlic in some cottage cheese or other goody that your chicken loves. You can also mix a teaspoon of fine garlic powder into a gallon of your chickens' drinking water, but that way, all your flock will be drinking the remedy. The garlic won't hurt them, but it may imbue your hens' eggs with a strong, distinct scent and/or flavor. If your sick chicken fails to make a comeback after a few days' worth of the garlic treatment, review *The Chicken Health Handbook,* by Gail Damerow, for more information.

> # Chick Chat
>
> Hens are female adult chickens. Pullets are female chickens less than one year old.
>
> ■ ■ ■ ■
>
> Buy baby chicks only from reputable feed stores on online hatcheries.

## Can chickens get frostbite?

Yes. Although chicken feet are quite resilient against frostbite, the wattles and comb are susceptible. So keep your chickens out of the wind and snow. The smaller the chicken, the more vigilant you need to be about keeping it warm in cold weather.

### Do chickens have good eyesight?

Yes. Sight and hearing are a chicken's two best senses (besides their sense of humor). They find tasty bugs and grubs by seeing them crawl around in the dirt, or hearing them shuffle under leaves or grass. Chickens don't have a great sense of smell or taste, which perhaps explains why they think worms and beetles are delicious.

### Do chickens make good pets?

Yes, absolutely! People keep all kinds of birds as pets: parakeets, canaries, cockatiels, parrots. Why not keep chickens? They are no more difficult to care for than any of these birds, or any other pets like dogs, cats, or fish. Keeping typical farm animals in a personal residence is nothing new — remember the miniature potbellied pig craze of the 1980s? Sadly, the attraction to pigs in residence went belly up a short time later. Very few true miniature pigs were sold to a pig-loving public. Most folks were in possession of a basic, generic pig that grew to 300 or more pounds within a year. The public reeled from Big Pig Shock. We cooled off toward farm animals as pets, and rightly so. Pigs, like most farm animals, didn't work out because they are simply too large and require too much care and space from a typical city dweller.

Yet a pet-loving public would not be deterred from taking chickens under their wing. Chickens aren't large. Chickens aren't difficult to care for and maintain. And chickens are cute. For these reasons, it was only a matter of time before we invited chickens into our city and suburban neighborhoods.

And while they do make great pets, we are talking outdoor pets, not indoors. Chickens cannot be housebroken. While my hens would love to come in to the house and watch "Animal Planet" on TV, I don't let them in because of their indiscriminate waste elimination habits (in other words, they poop anywhere, anytime, all the time). However, I heard of one chicken lover who kept her pet chicken indoors and controlled the relentless droppings by making the chicken wear a diaper (no lie!).

### Are chickens smelly?

This is an understandable concern in any community, particularly those where houses sit close together. The answer depends on the chickens' owner. Chicken coops that are not properly and adequately cleaned at least once a week will start to develop that barnyard smell. If you follow the simple guidelines for coop care described in chapter 7, you will be able to proudly say, "My chickens don't stink!"

## Are chickens dumb?

While chickens don't grow up to do quantum physics or govern small countries, they are not stupid. Their intelligence is relative to their species. They are, after all, birds. They aren't as smart as an African Grey parrot, but they hold an MBA when compared to a common finch. Chickens also do quite well in learning to respond to certain stimuli. My chickens have learned all manner of things. When they hear the back door open, they know to run to the coop gate to greet me. When I call "chick chick," they come running, because they know I call them over to personally hand out special food treats. Or when I tap my fingers on the picnic table, they jump up onto the tabletop looking for imaginary goodies.

## Can a chicken love you?

Yes — sort of. Although nature didn't equip them with the same capacity for affection as a cat or dog, chickens do show fondness, in their own way. Just walk outside with a dish of scratch or cottage cheese and watch the chickens *lovingly* run right up to you. Okay, so chickens love you for the food, but that's something, right? My chickens like to sit with me on the arms of my outdoor chairs, just hanging out and preening awhile. To me, that's love.

## Where's the best place to buy chickens?

At your local feed store or on-line from a reputable hatchery. You can also obtain all your henhouse equipment and supplies there. Most feed stores carry the most popular purebred chickens, such as Barred Rocks, Rhode Island and New Hampshire Reds, Australorps, and Orpingtons, and a few of the recently available commercial hybrids, like Red Sex Link and Black Sex Link chickens. If you are just starting to keep a small flock, you can get advice from the feed store employees or from the hatchery web site (usually great sources of basic chicken information), or you can buy one of the many books (like this one) they all have available.

On-line feed stores and hatcheries have a terrific variety of breeds available. The downside to ordering your chicks from an on-line breeder or hatchery is that sometimes a minimum order of chicks is required — between 10 and 50 — which far exceeds the number of chickens that residents are permitted to keep. This trend is gradually changing, as some hatcheries are responding to the resurgence of small backyard flocks. Some on-line hatcheries let you order just one chick in a breed, if desired. Unfortunately, ordering chicks by mail carries the morbid risk of finding one or more chicks dead after the ordeal of cross-country shipping. Reputable hatcheries have guarantees to prevent or resolve that occurrence.

### Can you keep chickens in urban areas?

After all this, you're still asking? City chickens are marvelous! Most city codes permit keeping several chickens in residential areas. They are easy-to-care-for pets that provide your family with fresh, organic eggs. They are part of the recycling and compost chain in your garden. And for you and your family, friends, and neighbors, chickens provide hours of fowl amusement!

### Do you need a permit keep chickens?

Depends on where you live and the number of hens you intend to keep. Municipal codes in some cities and towns do not require a permit unless you want to keep more than three chickens. To apply for a permit, you may need to obtain the signatures of all residents within a certain distance of your coop, which evidences their consent to your keeping a small flock of chickens. See chapter 4 for more information.

### Where can I learn more about chickens?

I've listed various print and Internet resources at the end of this book. Over the past few years, the best book I've seen about chickens (besides this one, of course) has been *Storey's Guide to Raising Chickens,* by Gail Damerow.

If you are like me, the more you learn about chickens, the more you'll want to know. One day, without your realizing it, your chicken illiteracy will have transformed into chicken enlightenment!

CHAPTER

4

# Chickens and the Law

WHEN THE TOPIC OF KEEPING CHICKENS in city or other residential areas comes up, you don't want to get on the wrong side of your neighbors or the law. Upset neighbors can make your life miserable; the law can force you to send your chickens packing. As a prospective chicken keeper, long before you crack open your first fresh backyard egg, crack open the books on your city's laws and ordinances regarding the keeping of chickens. Just as important is to poll all the neighbors adjacent to your property and get their "yes" vote on your feathered friends.

## Courtesy First: Talk to Your Neighbors

Don't invest in any coop materials and supplies — and especially not any chickens — until you have first checked in with your neighbors to gauge their warmth toward having a chicken coop near their home. It is the considerate thing to do. Even if you find that your local laws don't require a permit for your chickens, talk to your neighbors anyway. Include them in the process. After all, small flocks of chickens in a residential backyard may still be considered an eccentricity in your neighborhood (though this perception is changing). Your neighbors will appreciate your openness and courtesy about keeping what many consider to be strictly farm animals on a modest parcel adjacent to their home.

Assure your neighbors that the chickens won't be a nuisance to them or any other residents. Reassure them that you plan to be vigilant about henhouse maintenance and sanitation, and that odor will be minimal. Tell them that you intend to have only hens, no roosters. Most people are concerned about the noise accompanying a rooster. Advise your neighbors that you know roosters are too loud and rowdy to keep in close quarters and, on top of that, city law prohibits keeping them. Hens have their moments of excited clucking, most often after they've just laid a big egg. However, unlike roosters, most of the day chickens are quiet, softly clucking to themselves.

By including your neighbors up front, you not only avoid any potential misunderstandings later about your pet flock, but you also incidentally create new chicken lovers in your community. When I first got started building my chicken coop, neighbors and

**BIRD WORD**

Many cities permit keeping a few chickens on residential property.

■ ■ ■ ■

Some city codes are available on-line.

■ ■ ■ ■

Roosters are almost always prohibited within city limits.

their kids would occasionally stop by and check on the coop's progress. Some even wanted to participate in building the coop. Everybody was excited, asking nearly every day for weeks, "Is the coop finished?" and "Did you get the chickens yet?"

When you finally do bring the baby chicks home, everybody and their kids will come over to see their fuzzy new neighbors. The kids love holding the fragile, peeping chicks, and the grown-ups are reduced to whispered awe as the chicks are gently passed around. From the start, my backyard flock was actually the block flock. Everyone stopped in during the chicks' early days to cuddle them and marvel at their soft innocence and utter dependence. Everybody loves chicks. As folks keep coming over to visit, they will grow to love your chicks-grown-to-chickens. Some will want their own. Give them all the good advice you're learning here. And remember — tell them to talk to their neighbors first!

## NEST BOX NEWS
■ ■ ■ ■ ■

To maintain peace in your neighborhood, get the support of all adjacent neighbors before applying for a chicken-keeping permit.

## Chicken Codes

If you have checked in with all your adjacent neighbors about keeping chickens, and they all consent, be happy, but not too happy. Before moving in your imminent flock, you have one more step: You need to find out whether chickens are permitted in your community. You don't want to put time and energy into building a coop until you are certain you will be complying with all local laws and ordinances.

One way to find out whether chickens are permitted is to call the animal control department or health department in your town. Tell the employee you're put in touch with that you want to know if keeping chickens is legal where you live. If it is legal, ask if there are any specific restrictions or requirements related to chickens. Take notes. If the answer is not readily available, the city clerk may need time to obtain the correct response and will ask to call you back. Be patient!

Be sure to ask the employee to provide you with the code number and section numbers that pertain to keeping chickens in a residential neighborhood. After telephoning first, you can also visit the city department, city hall, or county clerk's office to review the code yourself. Make copies of the pages of city code that pertain to keeping chickens.

If you have access to the Internet, the quickest and easiest way to review the local chicken codes is through on-line research. Most cities and towns have a web site with links to the municipal code. If

*Poultry Tribune,* circa 1940.

you can't find the codes at the city's home page, you can access a database service, like American Legal Publishing at www.amlegal.com or Municipal Codes Corporation at www.municode.com. Both services provide the index and text of the municipal codes available for selected cities in most of the 50 states for free (and for even more cities for a fee). Once you are able to call up your city or town codes, keywords that generally bring up the relevant information are *chickens, poultry,* or *livestock. Livestock,* in legalese, means typical domesticated farm animals, like cows, goats, sheep, and chickens.

Municipal codes in different cities are all organized basically in the same fashion. Major topics relating to city legislation (such as animal control and health and sanitation) are called "chapters" or "titles" in the city codes. The chapters provide general information about the topic. Specific requirements or prohibitions relating to the general topic are organized into "parts," "sections," and "subsections" in the codes.

Whether on-line or at the city or town hall, if you have trouble pinpointing the law on chickens, ask a clerk for assistance. Also ask for assistance if you find *no law* relating to chickens. You might have missed something, so you should double-check before investing in your urban coop. If there isn't a law in the municipal code expressly prohibiting chickens, you can probably keep them. This is usually the case in towns and cities in the Midwest and South, where chickens aren't as removed from daily household life as they are on the densely cosmopolitan East and West Coasts.

Depending on where you live, your city code may allow residents to keep a certain number of hens (no roosters, remember?) without obtaining a permit. For instance, here in Portland, Oregon, our "chicken" code, found in City Code Section 13, subsection 13.050, states that no permit is required unless you want to keep a flock of more than three hens. You may be required to obtain an annual permit for your flock of three or fewer if the city has received nuisance or odor complaints from neighbors.

If you need a permit, obtain the form from the appropriate city department. You may have to pay a nominal one-time or annual fee to the county clerk's office. In some areas, issuance of a permit is conditional upon the written consent from all neighbors whose property adjoins yours or who are within a certain prescribed distance from your residence or the coop itself. The logic of the law: If the immediate community agrees to a neighbor's desire to keep several hens, it is less likely that complaints about the urban fowl will come up later.

Just because you have a permit to keep half a dozen hens doesn't mean you will always have the permit. If there any complaints about your chickens' behavior or if coop conditions necessitate a visit from a city inspector, you may be required to have annual inspections prior to obtaining each year's current permit. Also, permits can be revoked if your chickens cause any kind of nuisance, like excessive clucking or a foul-smelling coop, that results in repeat visits from a city inspector or other representative.

City codes often grant the permit not only contingent upon the informed, written consent of neighbors, but also with certain setback restrictions on the location of the coop in proximity to adjoining neighbors' property lines, doors, and windows. Some codes require that the coop be 20, 50, or 100 feet from the nearest neighbor's windows or doors. Not only are these setback requirements the law, they are the right and polite thing to do (I'm a stickler for courtesy). Set your coop as far away from your house as you can. Be sure you also set it as far as possible from your neighbors' homes. If you don't want to smell chicken coop in your kitchen, neither will your neighbors.

New suburban bedroom communities (a.k.a. planned unit developments, or PUDs) almost always prohibit keeping chickens in the neighborhood. These rigidly ruled communities like to keep themselves looking clean and sanitary. For a nominal homeowner association fee, a management company flunky will police your neighborhood and issue you a citation if you have more than five weeds on your property at any time. Needless to say, the PUDs have rather restrictive covenants about what homeowners may do with their homes. Such communities barely allow residents to keep dogs, much less chickens. Nonetheless, I'm hopeful. I believe that as more people start to keep small flocks of chickens in their gardens, the regulations of these PUDs will someday change to be more amenable to chickens in city and suburban neighborhoods. We should all have a right to fowl company in our gardens!

Finally, while the coop you build may be small enough not to necessitate a building permit, you should check with the building department or planning department where you live to make sure. Unless you are building an A-frame cabin on your standard-size city lot, you probably will not need a building permit.

After determining that a small flock of chickens in your residential yard is legal, obtaining the necessary permits, and getting your neighbors' consent, you can finally start to think about building your urban coop.

CHAPTER

5

# Building a Coop

UILDING A COOP CAN BE GREAT FUN. In this sort of project, the planning and design stage is just as fun — and as important — as the hammering and sawing. You need to address several preliminary questions before you strike the first nail. What sort of protection do your chickens need from the weather? How big does the coop need to be? Do you want the coop to be a fancy fairy-castle affair or a plain and sturdy pen? What type and quantities of supplies will you need? Do you have a place for the coop in your yard where it will comply with any existing building setback requirements? Building a chicken coop and keeping a small flock of chickens in your garden will be easy once you have answers to these questions.

## What's in a Coop?

The term *coop* refers to the entire hen habitat, which includes a chicken run and a henhouse. The chicken run is the outdoor portion of the coop, enclosed by chicken wire. The floor of the run is usually dirt and should be covered with gravel or absorbent materials, such as straw and cedar shavings. It can be enclosed with a roof or left open to the elements. Obviously, the chicken run and the chickens in it will stay cleaner and cooler if the run is roofed.

What do chickens do in the run? They hang out. Think of the run as a poultry living room or a chicken bar and grill. They eat and drink in the run, walk around, dig in the dirt, cluck among themselves. When they wake up in the morning, chickens head right out the hole in the henhouse door to the run to do the same thing they did the day before. And they love it!

The henhouse is a fully enclosed wood structure inside or adjoining the chicken run. Your hens sleep in the henhouse at night and lay eggs in it during the day. A small doorway in the henhouse opens into the chicken run. The floor of the henhouse should be raised off the ground on cement blocks or a solid cement pad.

Inside the henhouse, on opposite walls, are one or more perches and nest boxes. The perch is where the hens sleep, or roost, at night. Nest boxes are small, private cubicles where hens lay their eggs. Both perches and nest boxes should be firmly secured to the henhouse walls and elevated some distance from the floor.

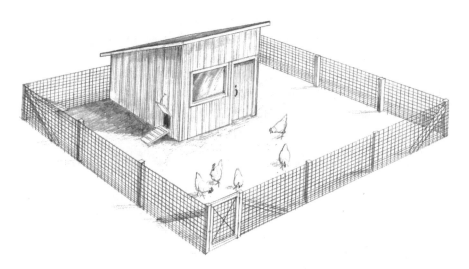

**The coop is bounded by a sturdy wire fence. The henhouse, located within the coop, is where the chickens sleep and lay eggs.**

# How Big?

How much space do chickens need? Per chicken, no less than 2 square feet (0.2 m²) in the henhouse and 4 square feet (0.4 m²) in the run. Halve those figures for bantams. A coop that is 10' x 4' x 5' (305 x 112 x 152 cm) with a henhouse that is 4' x 4' x 5' (122 x 122 x 152 cm) is a vast, comfortable estate for three chickens. In my neighborhood, I've seen smaller habitats that work, though they don't allow any extra wing room for the chickens. If you plan to keep your chickens in a smaller coop, you'll need to let them roam free in the garden a few times each week. They'll be a lot happier and healthier if they can stretch their legs from time to time.

If you have the room, give the hens extra space in the coop. If you don't give your chickens enough room, they will get anxious and start to stress out. Chicken anxiety creates coop chaos by manifesting in bad habits and compulsions, like pecking and biting one another, feather pulling, eating their own eggs, and cannibalism. For a happier place, give chickens their space! Then again, don't be impetuously motivated and build a coop big enough for 50 hens. Keep the coop at a reasonable scale for your lot and neighborhood. After all, we're talking about housing just a few chickens.

Another good idea is to make the coop and henhouse tall enough that you can walk in them. Crawling into your coop and henhouse to clean them is no picnic, no matter how much you like your chickens. Also, feeding, watering, and egg collection are easier to manage if you can walk inside the henhouse. You may want to install two doors into the henhouse — a chicken-sized door leading out into the run and a human-sized door at the back allowing you to enter the henhouse without having to walk through the run.

# To Roof or Not to Roof?

I recommend enclosing the coop completely, including a roof. By entirely enclosing the chicken coop, you protect your hens from predators like stray dogs, cats, raccoons, and occasional fly-by raptors. The roof can be made of netting, leaving the run open to the elements, or of a solid surface, such as plywood. A solid roof (made of plywood or tin, for example) will provide shade and keep rain off the chickens when they are out digging on a rainy, winter day.

If you're keeping bantam hens, definitely enclose the chicken run with some kind of roof; bantams have no problem catapulting themselves over a 6-foot (1.8 m) fence. Heavy breed hens, like the Girls, get only a few feet of altitude when they try to fly (but they do get a standing ovation for trying).

# Building in Cold or Hot Climates

The climate where you live will influence not only what breed of hens you get (see chapter 6) but also what kind of coop you build.

In colder climates, the henhouse should be equipped with a heat lamp for cold nights. It also should employ double-walled construction for additional insulation against the elements.

In warmer climates, the coop should encompass plenty of shaded areas. The henhouse should have windows that you can open on hot days for additional ventilation. You may want to design the

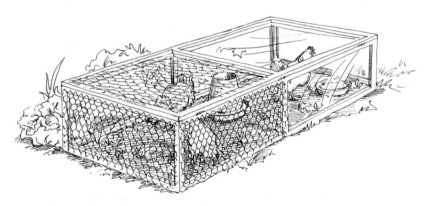

**A mobile chicken pen allows you to have your chickens "graze" different sections of your yard, tilling and composting the soil as they go.**

henhouse roof so that it's hinged to the frame instead of permanently affixed; you can raise the roof and prop it open for extra ventilation on especially hot, humid days. Another special option for a warm-weather coop might be an automatic watering system that ensures a continuous supply of fluids on the hottest days. If you're keeping large hens, install a box fan in the coop to help keep your big girls cool. How do you know if your chickens are feeling too hot? Well, if you're feeling too hot, chances are the chickens are, too.

A compact, warm-weather coop can be designed to be mobile. While mobile coops are most practical on larger parcels, they can also be used by suburban gardeners. The idea is to put the coop and chickens in one area, let them naturally compost and till the soil for a few weeks, then move the birds to another spot where the soil needs tilling and amending. This way, you (forgive the pun) get two birds in one shot — housing for your flock and year-round direct-soil composting.

# The Right Coop in the Right Place

*Location, location, location* is the guiding mantra for real estate, no matter what type of building — a simple restored cottage, a sprawling Victorian mansion, or a chicken coop. In the long run, location matters much more than appearance.

If local code specifies certain setback restrictions, put these into play in your yard. You know about how large the coop needs to be (see "How Big?" on page 53). Eliminate from the list of possible sites those that won't accommodate the coop without exceeding the setback boundaries. Depending on the code where you live, the boundary may be related to the nearest door, window, or property line of neighboring residents.

Common sense and courtesy dictate that you should locate the coop site as far away as you can from your windows and your neighbors' windows. By practicing coop etiquette in the planning stages, you prevent the likelihood that the scent of summer-warmed chicken droppings will waft into any nearby windows — including yours. As an added preventive measure for my own coop, I planted jasmine and honeysuckle on one side of the henhouse and run. Their lovely and scented blossoms upstage any unlikely possibility of a whiff of chicken coop on a breezy day.

Make sure the proposed chicken run gets adequate amounts of both sun and shade, so if you're not home during the day in summer, your chickens won't roast inside their coop. Also make sure that the proposed coop is not sited on ground that's lower than the rest of the yard. Otherwise, be ready to provide your hens with snorkeling equipment during rainy weather.

When you have settled on the most likely coop site, envision where you will put the door leading into the coop, making sure it will have plenty of room to swing out. You must give yourself access to the nest boxes (where hens lay their eggs), especially if you are building a short coop you can't enter standing up.

Once you find the right site in your yard, use string or spray paint to outline the general size of the coop area on the ground. Double-check to make sure the run has enough room for each hen. Check the paint or string boundaries once and then again. If you are confident that the coop is the right size, is in the right place, and falls within setback regulations, then you're ready to proceed with construction.

## Breaking Ground

Chickens don't care what their coop and henhouse look like, so long as their living arrangements are clean, dry, warm, easy to clean, and adequately ventilated. As you imagine your future coop, begin in your mind's eye with a basic structure made from wood — a shed, an A-frame cabin, or an enclosed lean-to. Then let your imagination run free, like a chicken in an unfenced clover pasture.

Supplies and equipment for a simple coop and henhouse can run as low as $100. A larger, roomier coop, like the one I built, might cost a few hundred dollars. All supplies are readily available at any well-stocked hardware store.

## THE FOUNDATION

To protect the floor of your henhouse from ground moisture, which leads to rot, you must keep it raised off the ground. Setting the henhouse on cement blocks will work. However, setting the house on a solid cement slab is better, because it insulates the birds from drafts and prevents incursions from tunneling rats. A cement slab can also serve as the actual floor of the henhouse; it's easy to scrape and sweep.

If you decide to go with the cement blocks, make sure that the floor of the henhouse, when set on the blocks, is level. You may have to dig out around the blocks and fiddle with their placement to achieve a level footing.

If you choose the cement slab instead, you can pour it yourself (if you know how) or have a local contractor do it for you.

## ENCLOSING THE RUN

The run can be constructed of chicken wire, chain-link fencing, or heavy-duty netting — anything that secures the hens away from outside dangers and predators, such as cars and neighborhood cats.

If you've ever fenced in a garden, you know that it's tiring, sweaty work. In the case of the chicken run, you'll need to make sure that the work is carefully done and the resulting fencing sturdy — you wouldn't want an animal throwing itself at an unsuspecting chicken to hit the netting and knock it down. To be installed properly, chain-link fencing generally requires the expertise of a professional. A fence of chicken wire or netting can be a do-it-yourself project. However, if you're uncertain of your carpentry skills, you may wish to have a contractor do the job for you. If you're willing to brave the challenge, read on.

The fencing will be supported by a series of anchor posts spaced 3 to 4 feet (0.9–1.2 m) apart. The anchor posts will also support the roof rafters, so they must be placed symmetrically — each anchor post must have a corresponding post

*Poultry Tribune*, circa 1940.

opposite it. Most of the coops I've seen use 2 × 4s as anchor posts. Cut the posts to the desired height, taking into account that they'll be set 8 to 12 inches (20–31 cm) in the ground. You may wish to have the posts on one side be 6 to 12 inches (15–31 cm) taller than the posts on the opposite side, with the posts on the two remaining sides (if any) sloping from one end to the other. The difference in height will cause the roof, when fit on the rafters, to slope, forcing rainwater to drain off in the direction of the lower side.

Use a post-hole digger (available for rent from most garden centers) to dig a hole for each post. Prepare a batch of quick-drying cement. Place a post in its hole and pour the cement into the hole. Hold a level along the side of the 2 × 4 to ensure that it is upright, and hold the post steady in place until the cement sets. (This job is easier done by two people — but quite amusing if done by one and watched by another.)

Once the anchor posts are secure, cut and fit roof rafters over the anchor posts, spanning the gap between each anchor post and the one opposite it.

It's usually easiest to staple the netting or wiring, attach the roof, and install a door and doorframe to the run *after* the henhouse has been framed. You'll have to look at your particular setup and decide the proper sequence of steps.

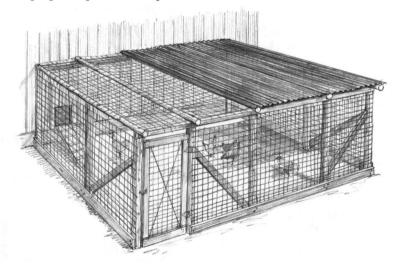

**A fully enclosed coop gives chickens freedom to roam without fear of predators.**

## ERECTING THE HENHOUSE

The henhouse should be weather-tight and draft-free for the winter yet able to provide adequate ventilation for hot summer days. You can reduce the possibility of drafts by siting the henhouse adjacent to a protective wall, a fence, or shrubs. If this isn't possible, create a windbreak by planting shrubs near the henhouse or putting up a wood fence on one or two sides of the coop area. If you keep your hens warm and dry in their run and henhouse, they'll be happy and lay lots of eggs.

We've already discussed a few ways to keep hens cool in hot weather, but they're important enough that I'll mention them again here. To ensure your henhouse is adequately ventilated during summer, plan for one or more of the following:

- A roof that can be raised
- Windows that open on opposite sides of the henhouse
- Screened doors that open on opposite sides of the henhouse

However, make sure these openings are constructed so that they shut and seal tightly for cold or rainy weather.

If you've poured a concrete slab foundation, then the floor of the henhouse is good to go. If you're going to set the henhouse on concrete blocks, you'll need to frame a floor from 2 x 4s and place a large solid sheet (plywood works well) over the framework. For more detailed instructions on foundation work and framing, check out *How to Build Small Barns and Outbuildings* by Monte Burch (see page 142).

If the henhouse was designed to adjoin the coop, you may have installed 2 x 4 framing for the henhouse walls at the same time you were putting in anchor posts for the chicken run. If the henhouse is a freestanding structure, you'll need to frame and roof it separately. If you don't want to do a lot of measuring and cutting, consider buying a premade shed for use as a henhouse; you will most likely be able to put it together easily and then have to make only minor modifications to make it suitable for chicken living quarters.

If you've never framed anything before, I would suggest that now is the time to solicit the help of a licensed carpenter.

If the coop is close to your house and an outdoor electrical plug, you'll be able to run extension cords out to the coop to power fans and heat lamps, when necessary. However, if your coop is located a good distance from your house, you may wish to hire an electrician to wire it. Contact the electrician early on in the construction process to find out at which stage of construction he or she would prefer to work.

Paint or seal all exterior wood. However, don't use paint or sealant inside the henhouse — hens nibble on everything!

## INDOOR AMENITIES

Inside the henhouse, secure a wooden dowel 2 inches (5 cm) in diameter to a wall, no less than 2 feet (61 cm) off the ground. Chickens always roost above the ground, for warmth and for safety. To the opposite wall, attach two or three nest boxes — little wood compartments — also about 2 feet (61 cm) off the ground. Chickens are quite secretive when laying eggs, and they like to find cozy, out of the way places to sit and lay. Make your nest boxes about the size of a shoebox stood on its narrow end, or perhaps a bit bigger for

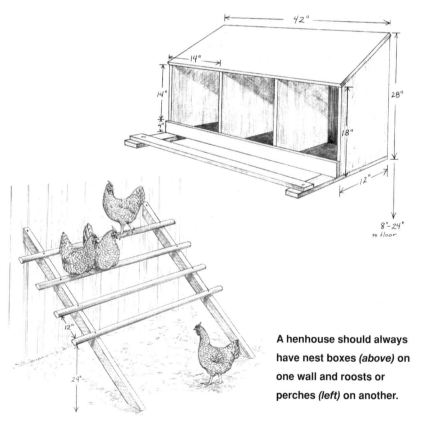

A henhouse should always have nest boxes *(above)* on one wall and roosts or perches *(left)* on another.

a "super-sized" hen. In colder regions, the back wall of the nest boxes should be against the inside wall of the henhouse, not against the wall exposed to the outside.

## FEEDERS AND WATERERS

Suspend a feeder and waterer from roof rafters of the run, about 6 inches (15 cm) off the ground. If you have an open-roofed coop, put the food and water dispensers in the henhouse, away from the perch. Make sure the dispensers are away from the coop and henhouse doors and all foot traffic to prevent spillage.

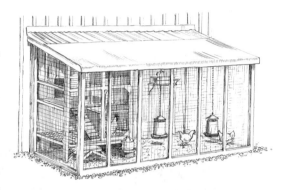

**Hang the feeder and waterer inside the coop's run.**

## FINISHING TOUCHES

The perfect coop depends on you. You can be a minimalist. Run chicken wire around some posts for the run. Set a big wooden crate on short posts or thick, heavy bricks, cut a square hole in it for a door, lay down a small plank as a door ramp, and rig up a perch and nest box inside for roosting and laying. It can be that simple.

Or be laboriously elaborate! Add decorative moldings to the henhouse and coop framing to invoke Victorian, Bavarian, or Art Deco styles. Paint the henhouse crazy colors and hang silly signs. Or paint a mural on it of a Mexican cantina, complete with faux shutters, climbing ivy, and a sombrero, and hang a painted sign over the henhouse door that says "Casa de Pollo" (that's "House of Chicken" for all you gringos). Use a French bistro mural theme and call your gaily painted henhouse the "Chez Poulet."

The only limits to your henhouse design are your creative genes and your wallet. So long as your hens are safe and sound inside with plenty of personal space, the coop and henhouse design are up to you. It takes vision, chutzpah, and a certain degree of wackiness to add a chicken coop to your city garden. Why should our chickens' coops be rudimentary when they can be extraordinary? The perfect coop is whatever you want it to look like.

# Building a Coop for the Girls

My coop started as a lean-to shaped like a parallelogram alongside the house. The site available for the entire coop was narrow but long. I sank ten 2 x 4 posts in the ground with a concrete base, leaving 5 feet (1.5 m) of each post aboveground for the roof to sit on.

My spouse and I framed the roof on a slant to run off rainwater and topped it with corrugated tin. Tin roofs sound lovely in the rain. Along the eaves of the roof, we installed a PVC-pipe gutter system to divert rainwater away from the coop and into the yard. To make the gutter, we simply snipped away one-third of the diameter of the pipe, leaving a U-shaped piece. (House gutters would have worked as well, but they're more expensive than PVC pipe.)

Next were the henhouse walls and doors. We built two human-size doors in the henhouse: one accessible from inside the coop and other at the rear, opening out into the feed and bedding storage area. We cut a small rectangular opening at the bottom of the front henhouse door so that the hens could enter and exit at will. We also installed a small slit of a door right behind the nest boxes. We call it the "egg door," because it allows us to collect the eggs without having to enter the henhouse.

We attached the roof on the henhouse with two large hinges at the rear of the roof, nearest to the house. We can lift the henhouse roof and prop it open during hot summer days for extra ventilation.

Finally, we poured a decorative concrete walkway to the coop from the patio and trimmed it with river rock for a serene effect and to encourage drainage away from the coop. We planted native ferns, ivy, and other durable creeping and shade-tolerant plants alongside the path. The several shrubs near the walkway are sturdy enough to be practically chicken proof.

Once or twice a year, I touch up the coop and henhouse with paint and make whatever small repairs are necessary. Mostly, I just enjoy the coop and its residents.

I used to look down on the coop from a rear window in my house, but I could see only a sliver of hen action from that vantage, not nearly enough for me. I made an offhand remark to my spouse about this small dilemma. Have I mentioned my spouse's nearly unhealthy obsession with electronics? Once word of my chicken-image deprivation was released, a solution was quickly announced. The chickens went wireless.

Wireless remote cameras, that is. We installed three of them in the henhouse and coop and then hooked them up to our TV. We can get up in the morning and turn on "Chick TV" to see what the Girls are doing, which is probably scratching and eating, but I guess I like watching that. If you love chickens as much as I do, you enjoy watching whatever they are doing. Believe me: To remote-view your chickens is to love them!

## Chick Chat

Design and build your chicken coop so it is as safe and secure from rats and predators as possible.

# Rats!

Most coops, no matter how ingenious and sturdy, won't keep out rats completely, regardless of whether they are slim, surreptitious barn-yard rats or fat, bold, slow city rats. Rats are expert diggers and tunnelers. If they smell chicken feed inside your coop, they will dig under the coop and pop out of a tunnel they have burrowed unerringly to right under the feeder.

I've tried subterranean concrete barriers. I've tried sinking lengths of fencing below ground. Neither worked very well to keep out the vermin. In my experience, there's no sure-fire way to keep rats out of your chicken coop if they're determined to get in. My advice: Be prepared for a long, continuous battle against rats.

I have found that rats are somewhat seasonal (at least up here in the Northwest). Rain in winter seems to bring them up out of the ground. Spring rat flings and resulting newborns bring rats forth like unwanted weeds in March and April. Things settle down during the summer and fall, as if the rats had packed up and gone to their summer cottages.

Don't be discouraged. There is a way to deal with rats: poison. Poison is tricky, however, because you have to be crafty enough to give the rats access to the poison and bait trap but careful enough to keep your hens and other neighborhood animals out of the traps' vicinity.

A few years ago, the city of Portland dealt with an increase in the city rat population by handing out "rats only" traps. These were plastic black bait boxes with a small hole big enough for a rat to fit into. The hole led through a short maze into the opposite end of the box containing rat poison. The rat ate the poison and left contented, then got a fatal stomachache a short time later. This trap set-up is quite effective, and I have never had a problem with one of my own critters getting into the bait box. Inquire at your local health department about these types of rat traps.

If your health department does not have bait traps available, you can build your own from scrap wood and nails. Construct a wooden box 1½ feet (46 cm) long and 1 foot (31 cm) wide with a hinged and clasping or locking lid. Cut a rat-sized hole in the narrow end of the box. (Remember, rats can squeeze into an opening one-fourth their size.) Insert two partitions inside the box to create a short maze to the bait and poison location at the other end of the box.

If you're setting a traditional "neck snapper" rodent trap, place it just outside the coop and henhouse at night. Set the trap inside a box just big enough for the trap and the rat. This way, there's no danger of your hens or other pets getting caught by the trap. Bait trap with peanut butter for a couple of nights without setting the trap. On the third or fourth night, set the trap with the bait the rat has come to expect. You will probably have a rat to dispose of in the morning when you check on your hens.

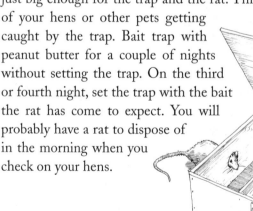

**A bait box made from scrap wood provides a safe way to offer poison to rats without inadvertently injuring other local critters.**

CHAPTER

6

# How to Pick a Chick

**W**ITH YOUR COOP PLANNED and construction under way, it's time to start thinking about your future pen of hens. Before you buy any chicks or chickens, it's important to educate yourself about all of the different breeds. By familiarizing yourself with the specific characteristics of the many chickens available, you will be more likely to pick chickens that will do well in your region and meet your needs for eggs and entertainment.

## Research

If you have access to a computer, start your research with the Internet. The World Wide Web hosts dozens of sites devoted to chickens, where people from all over the world share their chicken stories, tips, and pictures. A few of my favorite web sites are listed in the appendix, and hundreds of others I haven't listed are waiting for you to find them.

Many on-line hatcheries have chicken facts and links to yet more chicken sites. Some general information chicken web sites will lead you to great books covering every facet of chicken ownership, from urban chicken keeping to small farm chicken breeding. Before you know it, several hours will have passed on the Internet while you've been reading and thinking chicken.

*Poultry Tribune*, circa 1940.

If you don't have a computer in your home, go to your local library. Most libraries have computers with Internet access that you can use for free. If you are computer shy, read the books at the library to get you started. Great books about keeping chickens are available at all good bookstores, feed and supply stores, and yard and garden shops.

Picking chickens is as much as, if not more fun than, designing the coop. After reading about breeds, you can make informed decisions on which chickens you want to keep in your urban flock. Perhaps the color or pattern is the primary criterion for picking your chicken. Should you get big chickens or small ones? Maybe you want a mellow friendly bird, or one more inquisitive? Have several breeds (and styles) in mind when shopping for chicks. When you visit your local feed store in spring, it may not have all your first picks. The eclectic or rare breeds are almost always available from reputable Internet poultry hatcheries and poultry supply stores.

## Classifying Chickens

Purebred chickens are organized according to class, breed, and variety. *Class* is the broadest category of chickens, and it generally refers to the regions where the chickens were originally bred or with which they are traditionally associated. Chicken classes include Asiatic, American, Continental, Mediterranean, English, Oriental, and Miscellaneous.

Within each class of chickens are particular *breeds*. Chickens of the same breed have a similar body type, features, and markings. The breeds are basically divided into two categories. *Standard* breeds are medium to heavyweight chickens. *Bantam* breeds are smaller, lightweight chickens. The standard breeds tend to be mellower, while the bantams are known as "flighty" for two reasons: They are somewhat nervous all the time, and they can fly up to 6 feet (1.8 m) into the air, giving them plenty of clearance over most fences. Some breeds are excellent egg layers; others are well regarded for their meat. *Dual-purpose* breeds are known for both their meat and their eggs. Breeds with striking features or feather arrangements are sometimes known as *ornamental* breeds.

A breed can contain several *varieties.* A variety generally has a specific pattern or color of plumage. For example, Plymouth Rock is a breed in the American class that comes in several varieties, including a pattern known as barred (black and white herringbone) and colors like white, buff, blue, or silver. Wyandottes, one of the prettiest dual-purpose breeds, come in several varieties of solid colors as well as a vibrant silver-black or gold-black fish-scale pattern.

The *American Standard of Perfection,* published by the American Poultry Association, describes the ideal conformation and standards for each breed. *Show strains* are chickens whose physical type and traits conform to the breed standard. These supermodels of chic chickens travel nationwide to compete in poultry shows.

While the show strains in a breed are the upper-crust birds of chicken fancier society, *utility strains* are the blue-collar, working-class chickens within their breed. These chickens are purebred, but they possess certain imperfections that render them incapable of conforming to the breed standard required in poultry showing. These imperfections might be something as sinful as a narrow breast in a breed whose standard is a wide breast, muted or runny colors in a breed with a patterned plumage standard,

## NEST BOX NEWS
■ ■ ■ ■ ■
Each class of chickens contains several breeds, which themselves may contain several varieties with specific patterns or colors of plumage.

The hens of dual-purpose breeds tend to be tamer and less broody than hens of other breeds.

or misshapen beaks or feet. There's nothing really wrong with these birds — they're just not perfect supermodels! They are the "regular folks" of the chicken world. The birds you find for sale in local feed stores are usually utility strains of certain breeds. Though not up to show standards, these chickens have no problem laying eggs, digging compost into your garden, and making you laugh.

In addition to purebred chickens, you may also come across chicken *hybrids*. Most hybrids, also known as *crossbreds*, were developed by commercial poultry growers so that they could obtain the maximum amount of eggs and/or meat from a chicken. Hybrids tend to be plainer-looking than purebreds, and they can be quite nervous, a trait you may not want in your urban flock. An example of a popular commercial hybrid meat production chicken is the Cornish-Rock. This meaty bird was the result of an English Cornish being crossed with an American White Rock. Good egg-producing hybrids include Black Sex Link and Red Sex Link hens.

On page 69 is a partial listing of chicken classes, breeds, and varieties to get you familiar with the wonderful world of chickens.

While many standard breed chickens have a bantam equivalent, there are few true bantam breeds. The most popular of these are the Silkie, with its fine, white flowing plumage, and the Japanese, with its long, elegant tail feathers. Ah, so many chickens, so little time!

## Three Makes Company

Chickens are social creatures. They like company: their own company and yours. My chickens like it when I talk to them out the window, or when I sit and read with them in the yard. However, I can't always be with my chickens. That's why I have several — they keep each other company. No matter what breed or variety you decide on for your pet urban flock, don't get just one hen, because she

## Chicken Classes and Common Breeds

| Class | Breed | Colors/Patterns of Plumage | Avg. Weight of Hens |
|---|---|---|---|
| American | Jersey Giants | Black or white | 10 lb (4.5 kg) |
| | New Hampshire Red | Red | 6 lb (2.7 kg) |
| | Plymouth Rock | Barred, blue, buff, silver, or white | 6 lb (2.7 kg) |
| | Rhode Island Red | Dark reddish brown | 6 lb (2.7 kg) |
| | Wyandotte | Black, blue, buff, gold-laced, or silver-laced | 6 lb (2.7 kg) |
| English | Australorp | Black | 7 lb (3.2 kg) |
| | Cornish | Buff, white, or red-laced | 8 lb (3.6 kg) |
| | Orpington | Black, blue, buff, or white | 7 lb (3.2 kg) |
| | Dorking | Silver-gray, white, or red | 6 lb (2.7 kg) |
| Mediterranean | Ancona | Black with white dots | 4 lb (1.8 kg) |
| | Blue Andulusian | Slate blue/gray | 5 lb (2.3 kg) |
| | Sicilian Buttercup | Buff or black | 5 lb (2.3 kg) |
| Asiatic | Brahma | Light, dark, or buff | 10 lb (4.5 kg) |
| | Cochin | Black, blue, buff, silver-laced, or white | 9 lb (4.1 kg) |
| | Langshan | Black, blue, or white | 8 lb (3.6 kg) |
| Continental | Hamburg | Silver or gold | 4 lb (1.8 kg) |
| | Houdan | Black with white markings | 4 lb (1.8 kg) |
| | Polish | White, black, buff, or silver | 4 lb (1.8 kg) |
| Oriental | Phoenix | Silver or black | 4 lb (1.8 kg) |
| | Sumatra | Black or blue | 4 lb (1.8 kg) |
| Miscellaneous | Ameraucana | White, silver, brown/red, blue, or buff | 5 lb (2.3 kg) |
| | Araucana | White, black, or red | 5 lb (2.3 kg) |
| | Frizzle | Black, red, white, or buff; feathers curve outward instead of lying flat against the body | 4 lb (1.8 kg) |

will get lonely when you are not around to talk to her. Nothing is sadder than the mournful moaning of a bored, lonely chicken. Chickens can be quite vocal about their emotions.

## Climatic Considerations

When thinking about what chicken to get for your flock, keep in mind the climate where you live. Standard breeds don't tolerate extreme heat or humidity very well. Bantam chickens, by contrast, which weigh just 2 to 4 pounds (1–2 kg), have less body mass to cool off and can take the heat. On the flip side of the coin, standard breeds have ample insulation against the cold, while bantams in a cold coop will shiver themselves down to nothing.

If you live in a region with freezing cold winters (Northwest, Northeast, Midwest), pick chickens of the heavier standard breeds. In the Southwest, Southeast, and Deep South, where winters are milder and most summers are consistently hot and humid, heavy hens would do as well as a polar bear in a hot tub. Bantams or lighter medium-weight standard breeds are best for such temperate or subtropical climates.

## Good Breeds for a Backyard Flock

Having learned a little bit about breeds, you now have a better idea of what kind of chickens you may want. You already know that chickens come in two different sizes: bantam (small) and standard (large). Of course, besides the size differential, not all the chicken breeds are created equal. Some hens, like Plymouth Rocks and Rhode Island Reds, can weigh in at 6 to 7 pounds (3–4 kg), while others, like the

The Sebright *(left)* is a good example of the size of a bantam breed, and the Leghorn *(right)* of a standard breed.

Jersey Giant and the Brahma, tip the scales at a hefty 10 pounds (4.5 kg). A 10-pound chicken? That's the size of a small turkey! While such hefty chickens are the exception rather than the norm, they do exist. But before investing in one of these "big-boned" breeds, remember: The bigger the chicken, the more it eats. And chickens love to eat. If a low feed bill is important to you, pass on the big Brahma and Jersey Giant and consider a svelter breed, like a Rhode Island Red or Partridge Rock.

**BIRD WORD**

Good chicken breeds and varieties for urban flocks include chickens that are tame, quiet, hardy, and not broody.

Chickens come in a vast array of colors and feather patterns. They can be plain or fancy. They can have fluffy tufts on top of their heads, or have bare naked necks. They can have long slender legs, or legs so thick with feathers they look like waddling, fluffy pears. Some chickens have big, showy combs atop their heads, while others have just a short, blunt mohawk for a comb. Chickens are black, white, buff, brown, red, orange, gray. Chickens come with spots, checks, barring, or solid patterns. As with any species, it takes all kinds.

As I've mentioned several times, my Girls include a Barred Plymouth Rock (Zsa Zsa), a Rhode Island Red (Lucy), and an Australorp (Whoopee). Out of the three, Lucy is the best layer, despite her breed reputation as a moderate laying dual-purpose breed. Lucy gives me a medium-sized, speckled brown egg nearly every day.

The next best layer in my flock is the Barred Plymouth Rock. Zsa Zsa lays four to five large or extra-large eggs on me a week. If Zsa Zsa misses a day or two of consecutive laying, she will leave a huge egg the next session to make up for it — and not infrequently it's a double yolker!

The Australorp (Whoopee), the biggest hen in the flock, lays eggs the most sporadically. Australorps are reputed to be prolific layers, but my Whoopee is not one of those. Thank goodness Whoopee is fun and good looking, because she sure is lacking in the egg-laying department!

I mentioned the Girls' laying habits for a reason. Individuality counts. You may pick one breed because it's reputed to be an Olympic-quality layer and another breed for its good looks, but chickens, like people, are not all created equal. A chicken you choose for plumage may end up being a veritable egg machine, while the reputed layer drops an occasional egg as little more than an after-thought to a day of grubbing and clucking. Since garden flocks are usually kept as much for fresh backyard eggs as their fowl company, I've listed what are considered the most dependable egg-laying hens in the table on page 73. These are all light- to medium-weight hens that would do well in most any climate (or what I otherwise refer to as the All-Season Hen Collection). Most of these hens are friendly, quiet, and not overly active. All are dependable egg layers, but none is reputed to be broody (that is, to have a tendency to sit on a clutch of eggs to try to hatch them). Brooding is a rather moot endeavor for the urban hen, as she has no rooster to fertilize the eggs. Brooding is also an eggless time for you, as the hen will not lay new eggs while being broody.

## Where to Get Chicks

Why get chicks instead of chickens? To begin, mature chickens aren't usually available for sale in the local feed stores or from on-line hatcheries. If you can find one for sale, a full-grown hen will cost you about $20 (compared to a baby chick that costs about a buck). After folks have spent several months of time and chicken feed plumping up their hens into maturity, they are not very likely to let them go (would you?).

However, if you really want a head start on hens, pullets that are three to five months old are sometimes sold at feed stores for around $10 each. For my money, I prefer getting a baby chick to raise myself. This way, you get the enjoyment of watching a chick grow up, and the chick knows you all its life. A hen you've raised from a chick is apt to be much tamer than an older pullet or young hen you would purchase. Plus, you have no idea how a mature chicken was handled before you brought it home. If you want the bird to be friendly, accustomed to being handled, and as well-mannered as nature will allow it to be, you'd do best to raise it yourself from chickdom.

# Good Egg-Laying Hens for Backyard Coops

| Breed | Eggshell Color | Hen Body Weight | Comments |
|---|---|---|---|
| Ameraucana | Blue to army green | 5 lb (2.3 kg) | Good layer. Cautious. |
| Araucana | Blue to army green | 5 lb (2.3 kg) | Good layer. Cautious. |
| Australorp | Dark brown | 7 lb (3.2 kg) | Good layer. Cautious. Extra hardy in cold weather. |
| Black Minorca | White | 5 lb (2.3 kg) | Good layer of white eggs. Cautious. |
| Black Sex Link | Dark brown | 5 lb (2.3 kg) | Great hybrid layer. Friendly disposition. |
| Buff Orpington | Medium brown | 7 lb (3.2 kg) | Great layer. Friendly disposition. Hardy in cold weather. |
| Leghorn | White | 5 lb (2.3 kg) | Great layer. Friendly disposition. |
| New Hampshire Red | Medium brown | 6 lb (2.7 kg) | Cautious and calm. |
| Plymouth Rock | Light brown | 6 lb (2.7 kg) | Great layer. Friendly and curious. |
| Red Sex Link | Medium brown | 5 lb (2.3 kg) | Good hybrid layer. Friendly disposition. |
| Rhode Island Red | Medium brown | 6 lb (2.7 kg) | Great layer. Friendly and calm. |
| Silver-Laced Wyandotte | Medium brown | 6 lb (2.7 kg) | Good layer. Very attractive plumage. Curious and calm. |
| White Wyandotte | Medium brown | 6 lb (2.7 kg) | Good layer and hardy in cold weather. |

If you want chicks, get chicks that are already born. Don't get eggs and try to hatch your own chicks. It's too involved a venture and requires more time and attention than you probably have to give. Hatching chicks is more appropriate as a science project for kids — after all, they have lots of free time on their little hands. Plus, if you incubate and hatch your own chicks, you'll have the problem of sexing them.

*Sexing* is the professional poultry grower's term for the process of identifying chick gender. Chicks at the feed store are divided by breed, then subdivided into "sexed" categories, while a batch of unsexed chicks is identified as a "straight run" (includes both female and male chicks, as gender has not been differentiated). If you hatch

your own chicks, you'll get both male and female chicks, but you won't know which is which until the roosters start crowing. Get girl (a.k.a. sexed) chicks that are a day or two old. By getting them so young, you will bond with them just as well as if you'd hatched them yourself.

Call around several feed stores in your area to see what breeds they have on hand. If you can stand the suspense, visit a couple of the stores prior to purchasing a chick to see who has the best selection and cleanest habitats. Ask your feed store if the chicken breeder who supplies them with chicks is a member of the National Poultry Improvement Plant (NPIP). This organization promotes the responsible breeding of healthy poultry.

Picking a chick out of a full brooder is like trying to pick one fish out of an aquarium tank containing four dozen. Yet if you derive pleasure from choosing your own chick, even though she looks exactly like all the others, then the feed store is your place for future chicks.

The downside of buying your chicks at the local feed store is that it may not have eclectic or rare breeds. For this reason, more often than not, be prepared to leave with a healthy, perky baby chick, no matter what the breed.

If you are unwilling to settle for a backup and absolutely have to have that Gold-Laced Wyandotte you've read so much about, then turn on your computer and shop at some of the reputable, regional hatcheries. These professional breeders raise lots of chickens, and they raise chicks all year long. If you can't wait to get started, you can order chicks immediately over the Internet when you start designing and building your coop. While the benefit of shopping at year-round on-line chick hatcheries is having more breed choices, the downside is that many such hatcheries require minimum orders of 25 to 50 chicks (though this trend is changing as more pet chickens hatch in yards across America each day). No matter how well you plead your case to your neighbors or city representative, they are not likely to concede to a pet flock numbering in the dozens. Also, with on-line mail order chicks, there is always the unlikely though distressing possibility of finding a couple of dead chicks in the delivery box. Reputable on-line hatcheries guarantee a live chick to your doorstep. However, be prepared for the possibility of a fatality, and be extra happy when all the chicks you've ordered arrive live and peeping.

## How to Pick Chicks

First rule with picking your chicks: no boys — they become roosters, which you can't keep in a neighborhood yard. Pick chicks whose sex has been determined (known as *sexed* chicks) rather than chicks whose gender will be a surprise *(straight run)*. If you were raising chickens for meat, you would get straight-run chicks, as you would not care whether you got roosters or hens since you plan to eat them anyway. However, if you want eggs, you need hens, so get female chicks only. A sexed chick costs about 30¢ more than a straight-run

*Poultry Tribune*, circa 1940.

*Poultry Tribune,* circa 1940.

chick, but it's worth it. For a little over a quarter, there is a 90 to 95 percent breeder guarantee that the chicken will be a female.

When picking baby chicks, avoid chicks that look unquestionably sick, have visible deformities (bent beaks, twisted legs, watery eyes), or are listless, weak, and unresponsive. When you get to the feed store, you may be staring into a large box containing one hundred baby chicks. Spend some time trying to focus on a few chicks so you can visually inspect them. Ask the store employees if you can pick up a couple chicks to take a closer look at them. I've always spent an hour or more staring at baby chicks to pick and bring home. By carefully examining my chicks and selecting only the healthiest looking birds, I have never had a single chick mortality.

Be warned: Baby chicks are very cute. Resist the temptation to get more chicks than you need. Chicken farmers sometimes recommend getting 25 percent more chicks than planned because chick mortality is not uncommon when acquiring large numbers. This formula is unnecessary to the keeper of small flocks. If you pick healthy chicks to start with and take care of them — keep them warm, provide plenty of fresh water and food, and change bedding frequently — you shouldn't have any fatalities in your fledgling flock. Pick only the number of chicks you ultimately want in your flock, and take great care of them!

CHAPTER

7

# Chicken Care

**S**O, YOU PICKED OUT YOUR CHICKS. Now they're home with you, still bundled inside a dark cardboard box punched with ragged little breathing holes, frantically peeping. They want out — now! What do you do?

## The Care of Little 'Uns

Well, what you should have done before bringing the little ones home was set up their living quarters. No, not that spacious coop you're building or have finished building out in the backyard. You need something smaller and cozier for the chicks until they are mature enough to tolerate being kept outside. You need a brooder.

A brooder is a wire cage or some other type of ventilated box equipped with an overhanging light source for warmth. Some folks use old aquariums or wood boxes, but I prefer a large wire cage specifically made for the purpose of raising a few baby chicks. These

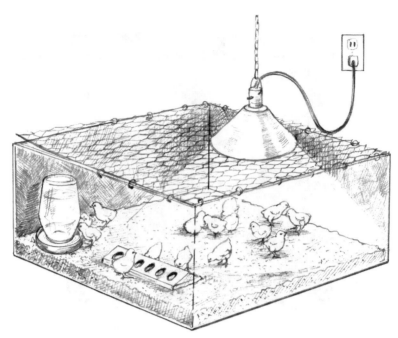

A brooder should provide water, food, warmth, quiet, and security for your new chicks.

cages or brooders are available at your local or on-line hatchery or feed store. Keep the brooder on a table or countertop in your basement, garage, or spare room while your chicks occupy it. You want the baby chicks close by to monitor them.

Ensure that no drafts will disturb the brooder. A great quick windbreak for a brooder can be as simple as newspaper folded lengthwise into 4-inch (10 cm) strips and taped around the bottom of the cage.

A typical brooder cage has a wire floor and sits on top of a removable metal pan for easy cleaning. Lift the cage, line the pan with newspaper, and replace the cage atop the pan. You will need to remove the soiled newspaper and replace it with fresh newsprint at least once daily, or else your chicks will start to stink. Chicks are, and remain throughout their lifetimes, prolific poopers. I always clean my chicks' brooder twice daily; in the morning after the chicks have had a long night to themselves to sleep and poop, and before my bedtime, so the chicks can rest in clean premises. Just because you're raising chicks doesn't mean you want your spare room, garage, or basement to smell like a barnyard!

## BEDDING

Put some old rags, towels, or socks on the cage floor. Chicks don't roost on a perch the first two or three weeks of their life. They fall asleep on the floor where they are standing, eating, or pooping. The rags give them something soft to fall asleep on. Use rags you won't mind throwing out. Once the chicks sleep (and, of course, poop) on them, you'll want to throw them away and replace them with fresh sleeping rags daily. Make sure the rags don't have loose threads on them or the chicks will try to ingest them, which can harm the chicks. I found that old socks are best to use for chick bedding — they have few loose threads and fibers that the chicks can pull on and consume.

## NEST BOX NEWS
■ ■ ■ ■ ■

Chicks live in their brooder, a roomy cage equipped with a heat lamp, until they are fully feathered. They are now called *pullets;* when they are one year old; they become *hens.*

After the chicks are a month old, you can place cedar shavings on the bottom of the brooder cage. Do not use cedar shavings before that first month is over, however. Baby chicks don't know what's what yet — in the course of experimental tasting, they will try to eat the shavings. If they do ingest wood shavings, their digestive system may become blocked up (known as *pasting up),* and the chicks may become very ill and die.

## TEMPERATURE

Suspend a heat lamp 6 to 8 inches (15–20 cm) from the top of the cage. The temperature in the cage should start out between 90° and 95° F and should be decreased by five degrees each week for the next five to six weeks. Put a thermometer on the cage floor to determine whether to move the heat lamp higher (to lower the temperature) or lower (to raise the temperature).

Don't hang the lamp dead center over the cage, but over to one side. You need to give the chicks room to escape the lamp if they are feeling too hot. You will want to regularly observe and regulate the heat so that you don't accidentally roast your chicks. If the chicks are always huddled together directly under the lamp, the brooder temperature is too cold. If the chicks stay as far away from the lamp as they can, clinging to the walls on the opposite side of the brooder, the temperature is too warm.

## WATER AND FEED

Your chicks need to have plenty of cool, fresh water to drink. Put the chicks' water dispenser in the cage over to one corner, away from the direct path of the heat lamp. If the water in the dispenser becomes too hot, the chicks will not drink. Chicks' fragile physiques are susceptible to immediate dehydration without access to fresh, cool drinking water. Discard the water from the dispenser and refill it with clean water twice daily. The chicks poop everywhere, all the time, and they make no exceptions for their watering tray.

Place a chick feeder in the cage. The feeder is usually a stainless-steel feeding dish, though plastic feeders are available (plastic is a bit easier to clean). The feeders can be either a shallow round dish with a top cover containing several half dollar–size holes, or a narrow trough topped lengthwise and center with a rod that turns in place. The purpose of the holes in the round feeder and the teetering rod in the trough feeder is to keep the chicks out of their food dish. Without the protective top and rod, the chicks, not knowing any better, would stand in their dishes and poop and sleep.

A chick feeder gives chicks access to their feed while at the same time keeping them from getting *into* the feed.

What goes in the chick feeder? Chick feed, of course. Chick feed has a unique nutritional makeup designed just for growing chicks. Chick feed is ground up so that it's easy for the chicks to eat and digest. It comes in two forms: "mash" and "crumble." Mash and crumble each look like their name implies. Commercial prepared chick feeds, which are available at feed stores, have all of the necessary nutrients needed by your chick flock.

*Poultry Tribune*, circa 1940.

Chick feed comes in two varieties: medicated and nonmedicated. Medicated feed prevents chick coccidiosis and is essential in larger farm flocks of chickens. However, with only three or four chicks, you have control over the cleanliness of their habitat. Disease in your small flock is not as likely to occur as in larger, farm-size flocks. I've always raised my city chicks on nonmedicated feed and have never lost a chick to sickness.

## ROOSTING PRACTICE

When your chicks are about three weeks old, install a small perch or dowel into one end of the brooder. Your fledglings need to practice roosting much like a kid needs to ride a tricycle before trying out a bike. Placing a dowel into the brooder early in their lives encourages the chicks to give roosting a chance. If you see that the chicks aren't getting the idea to jump up on the perch themselves, give them a hand. Pick them up and hold them over the perch until they grip it. Gently let go when they do. They will fall over a few times, but like a kid on a bike, once they learn how to roost, they'll never forget. And there's nothing cuter than several-weeks-old chicks, still all fuzzy and peeping, clustered on the perch together!

## FROM BROODER TO COOP

Put the chicks in the brooder as soon as you bring them home. Promptly give them water to drink. Initially, mix a little sugar (1 teaspoon per quart) into the water. This mixture gives them instant energy and helps reduce their stress level (imagine being locked up and bounced around in a dark cardboard box for a while). A little serving of sugared water is especially important if you have received your chicks by mail, as they will be thirsty and stressed from their long journey. Once your chicks have had some time to adjust, provide them with fresh water without adding sugar. The sweetened water is a one-time serving to the chicks and should not be continued after the first watering.

The first day and night with the chicks will be magical. Well, it's magical for the people; I imagine the chicks have a different take on the experience. They are frightened, awestruck, and totally dependent on you. They are fuzzy, clumsy, and curious. They are full of life but weigh little more than a heavy paper napkin.

Your new chicks will look fragile and wobbly. Their tiny claws will fall through the holes of the wire flooring. Don't worry — chicks are actually tougher than they look. Soon they'll get stronger and more accustomed to their new home. In just a day, the baby chicks will be running from side to side and over the wire flooring without losing a step. They will eat, drink, and sleep like old pros.

And they will peep. They peep all the time like chatty canaries on caffeine. They peep when they eat. They peep when they poop. Peep when other chicks are peeping. Peep when no one else is peeping. It is a veritable peep show (bad peeping pun indeed). They stop peeping only when they sleep, which they do suddenly without warning. Sleeping comes naturally and abruptly to chicks. To sleep, the baby chicks simply fall down wherever they are standing and peeping, and they close their tiny chick eyes.

I'll never forget the first time I held a baby chick. It was — what else — magical. The chick got really warm in my cupped hands and fell asleep with her tiny head resting on my fingertips. The first time I saw a chick asleep in the cage, I had such a scare. I thought she was dead. She was lying face down, wings slightly opened and splayed away from her body, looking lifeless. I tentatively touched her through the wire and she popped up, peeping, and ran to the food tray while pooping. Whew. If you see your chicks lying this way, don't panic. They may scare you to death the first time you see them lying there on the bottom of the brooder like a cottonball carcass, but they are probably just taking a quick nap.

Get used to seeing your chicks sprawled facedown in the poultry power nap position. They spend most of their first few weeks either sleeping or eating. The abundant rest and food fuels the chicks' physique through an amazing growth spurt. They'll go from a few ounces to a few pounds by the time they're eight to fourteen weeks old.

When can the chicks move out to the coop? Depending on their breed and variety, chicks will be ready to begin their pullet rite of passage out of the brooder and into the coop at anywhere from three to four months of age. The basic rule is: Wait until your pullets have all their feathers (that is, they're fully feathered) before moving them out to the coop.

The move to the coop should be undertaken gradually. Think of young pullets as greenhouse seedlings. Seedlings raised indoors are never brought out and immediately thrown into the soil. Instead, they are gradually acclimated to the outdoors over a brief period (known in gardening parlance as *hardening off*). It's the same for your young chickens. When the weather has become consistently warm, take the birds outside in their brooder. Leave the brooder in the henhouse for a few hours, then bring it back inside for the night. Repeat the next two days. On days three through six, open the brooder and let the pullets wander freely around their new digs, but continue to bring them in for the night.

After a week of this careful treatment, your pullets should be accustomed to the temperature and décor change in their habitat. Having had a chance to scratch in the dirt and eat bugs, they'll be itching to move into their new abode. Let them.

## Sick Chicks

Nobody wants a sick chick. Once a baby chick is sick, its chances for survival are not as good as if it had remained healthy through its chickhood. The best cure is prevention. First, pick chicks that aren't already sick. See chapter 6 for more information about that. Second, keep the cage clean, provide the chicks with fresh water twice a day, and keep an eye on their behavior. I bring lawn chairs into the basement where my chicks live and sit and watch them a little bit each day. By getting accustomed to your chicks' looks and habits, you are more likely to become aware of anything that doesn't look right.

*Poultry Tribune,* circa 1940.

## SIGNS OF DISTRESS

Possible signs of discomfort or illness in chicks may include watery eyes, waterier-than-usual droppings, listless behavior, and not eating or drinking. Chicks (and chickens) are susceptible to waste-borne disease and bacteria. If you are really watching your chicks, you'll notice that they walk around their cage pecking at and tasting big bites of their own droppings (remember, chickens do not have good senses of smell or taste). Keep the cage very clean to decrease the chicks' chances of consuming their own waste.

By watching your chicks carefully each day, you can prevent problems before they happen. An uncared for, unwatched chick could die from simply pasting up (see below). How will you know if your chicks are unhappy for some reason? You will hear them. They will peep and cheep more incessantly than they usually do. They may have run out of water or food (they never should). They may be too cold or too hot. Perhaps they are pasting up or just not feeling well. Watch your chicks diligently the first few weeks of their lives. They are pretty much helpless and rely totally on you for their food, water, warmth, and health.

If your chicks are ill and home remedies are not helping, you can call the feed store where you purchased them for some advice, or contact a local avian veterinary practitioner.

## PASTING UP: THE MOST COMMON CHICK ILLNESS

Pasting up occurs when a chick's droppings cluster up and adhere to its behind (the vent), preventing the chick from passing new droppings. It's the chicken version of constipation. A chick can paste up if it eats cedar shavings (a good reason not to use them for bedding) or if it doesn't have enough water to drink (chicks and chickens drink surprisingly large quantities of water).

If you see that a chick is pasting up, pick it up and use a damp, warm washcloth to gently remove the material from the chick's rear end. Try not to get the chick too wet with the washcloth or it can catch draft and a cold.

Watch your chicks daily to make sure they are clean and happy. Pasting up can cause your chick discomfort or, if not promptly noticed, death.

# Feeders and Waterers

When your chicks start to resemble fat, feathered pigeons, they are no longer chicks but pullets. They will be pullets until they are one year old. Then they'll be hens. Pullets and laying hens have different feed needs than do chicks. You can get all the feeds necessary for your chicks, pullets, and laying hens at your local feed store.

The bigger chickens are, the more feed they eat. Those little chick feeders aren't going to do the job of feeding your ravenous pullets and hens. To keep food plentiful and always available, use a galvanized stainless-steel cylindrical feeder. These can hold several pounds of feed. Suspend the feeder from the coop or henhouse rafters so its lip is about even with the flat of your chickens' backs when they are standing. Keeping the feeder raised helps keep the food in the dispensing tray clean.

You will also need a larger waterer for your chickens. Chickens are thirsty birds. How thirsty? The Girls can easily go through a gallon a day. They drink even more on hot days. Install a 3- or 5-gallon (11 or 19 L) plastic or galvanized steel waterer in the coop or henhouse. As with the feeder, suspend the waterer so that it is as high as your chickens' backs. Keep the water dispenser in a shady area to keep the water cool and fresher longer. Check the water level in the waterer every day to make sure your chickens haven't tilted it and caused it to leak and spill out. Refill or refresh the water as necessary.

## FEEDING PULLETS AND HENS

Pullet feed is basically hen feed without any calcium additives. Calcium makes for strong eggshells. Give your chickens pullet feed

until they are about four to five months old. At about this time, the birds are fully feathered, their butts are really fuzzy, and their legs and hips seem to spread farther apart so that they have a bell-bottomed shape. When your hens' bottoms start spreading, it means your girls are near egg-laying time.

As your pullets transition into henhood, gradually mix in hen laying feed (the one fortified with calcium) with the remainder of the pullet feed. The laying feed is made up of larger pellets than pullet or chick feed. Give your hens time to get used to this — at least two weeks — by mixing the adult laying feed into the remaining pullet feed in increasing increments until all the pullet feed is gone. By that time, your hens will have laid their first eggs.

Once your chicks have become pullets, introduce grit into their diet. Grit is gravel or small rocks pullets and hens need to eat, along with their chicken feed. Why? Because chickens don't have teeth. Think of grit as the only "teeth" the chicken has to chew its food. When a chicken eats, food goes into the crop first. The crop is like a chicken's initial stomach — it doesn't digest food, but merely prepares it for its imminent journey down the gullet and into the gizzard.

The gizzard is a chicken's muscular "second stomach." Grit joins ingested food in a chicken's gizzard. While the gizzard undulates like a flexing and relaxing muscle, the grit gets tossed around with the food, which gradually mashes up all the food for subsequent digestion. Without grit freely available to your hens, they will not digest their food properly and they may become ill.

Keep about a small cup of grit available at all times in a sturdy container that won't tip over. I use a clean tuna or cat food can posted on top of a wood stake in the ground. (I remove the label, punch a hole in the bottom of the can, and screw it into the stake.)

## STORING FEED

Keep chicken feed clean by storing it in a large, resealable plastic container. I use two large trash barrels: one with the open bags of feed and scratch stored inside, the other for the unopened bags of feed and scratch. Storing the chicken feed and snacks in water-tight, vermin-resistant containers lets me keep the food near the hungry hens, out back behind their coop.

## TASTY TREATS

If you really love your chickens, give them scratch. Scratch is a mixture of cracked grains and corn. It's sweet, fattening, and delicious — like caramel corn for chickens. They go crazy for it.

Scratch is a good snack for chickens, but it should never be substituted for the daily laying feed. Scratch contains a lot of corn, which is high in fat and low in protein. Your hens need at least 16 percent protein in their diet each day, which they'll get in the prepared hen feed, but not solely from scratch.

Since scratch is high in fat, I dish it out to the Girls before they roost in the evenings. I tend to give them a little more scratch in winter, so they'll have something in their gizzards to keep them warm on our cold Northwest nights. Feeding your hens too much scratch will make them fat — too fat. Plump hens are happy hens, but obese hens are susceptible to health problems.

When feeding your chicken, think in terms of variety. A chicken is what a chicken eats. Would you be healthy if you ate nothing but pepperoni pizza for breakfast, lunch, and dinner? Happy perhaps, but not healthy. It's the same with chickens. In addition to prepared chicken feed and occasional scratch snacks, they

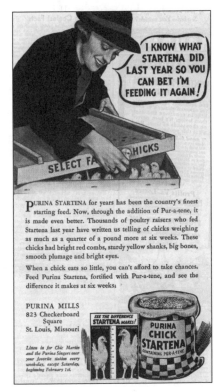

Poultry Tribune, circa 1940.

## Chick Chat

Feed your chicks not only chicken feed but also greens, fruits, and vegetables. They love the variety, and it will make their eggs rich and luscious.

need some variety in their diet. They especially need greens and vegetables. If you can't let them out in the yard often enough to nibble on your lawn, or if you have no lawn, toss some greens into the coop every day. Chickens can have even wilted greens and produce, so long as they are not spoiled.

I like to give my hens some fresh greens each day, together with a few pieces of fruits or vegetables. When lettuce and corn are on sale, I always get extra for the Girls. I also am fortunate to have a wonderful neighbor who works in a grocery produce department and brings the Girls tasty, fresh, and otherwise wasted greens and vegetables.

Chickens love to eat breads and starches, like rice and pasta, but take it easy on the portions you give them. Excessive amounts of grain products can make your poultry portly. Small bits of bread treats are fine. For instance, I don't like eating the heels on a sliced loaf of bread. Guess what? The chickens aren't as fickle as I am — they love the bread heels! When I have leftover rice or pasta that does not merit storing in the fridge, I toss it into the coop with the greens and vegetables.

When I want to really give my chickens a savory treat, or when I need to bribe them back into their coop, I bring out the heavy artillery: cottage cheese. I swear, the Girls would hold their eggs ransom for that cottage cheese if they could. And it's not only a proven bribe but also is full of vitamins and is a great source of protein. Once a week, I'll put out about a half-cup of low-fat, no-salt cottage cheese for the Girls. Why is low- or no-salt cottage cheese best? Because chickens already get whatever salt they need in their prepared feed. The Girls are on prepared feed because it has exactly what they need for optimum and balanced nutrition. Too much salt in their diet will not only make chickens thirstier than they already are, but it can also affect their laying and health (nobody wants a hypertensive hen!).

## Good Food for Chickens

- Lettuce
- Greens (spinach, chard)
- Tomatoes
- Corn (fresh, no butter)
- Bread and pasta (in moderation)
- Cottage cheese (low-fat, no salt)

## Bad Food for Chickens

- Spicy food
- Rotten food
- Sour food
- Potato chips (or any high-fat, high-sodium, deep-fried junk food)
- Cabbage (makes eggs stinky)
- Garlic (also makes eggs stinky)
- Raw potato peels (chickens have trouble digesting them)

# Eggs, Eggs, Eggs!

The coop is built. Your darling chicks have grown into pullets, which have matured into hens. The hens have been eating and drinking like queens. In fact, they are starting to look like King Henry VIII on chicken legs. So what's next? The moment we've all been waiting for — the eggs!

Eggs for cakes, cookies, quiche. Eggs for breakfast, lunch, and dinner. So many eggs that I have to give them away to grateful relatives, friends, and neighbors. Everyone I know was as excited about those first eggs as I was. In a funny way, eggs seem to bring people together. It did in my family and circle of friends, and it will for you and yours.

Once your hens start laying, people by the dozen come out of the woodwork, humbly pleading for those fresh, backyard jewels. Walk a balanced line so you don't deny anyone your backyard eggs, but do not overindulge anyone, either. While true largesse is giving away one or two dozen eggs at a time, dole out those gorgeous, yolky gems frugally, a half dozen at a time. This is not to be stingy, because you will have more than enough eggs for yourself even if you give half of them away. The parsimonious passing out of the eggs as a sort of continuing ransom will ensure a steady stream of egg-craving friends and visitors for your chickens.

When the Girls were about five months old, an informal "egg watch" began and was ongoing nearly 24 hours a day until the first egg was dropped. The phone would ring. The dog would bark as neighbor after neighbor stopped at the front door. Over and over, the question: "So . . . uh, any eggs yet?" When one of the Girls laid an egg the first time, everyone got a phone call (known in our family as the Egg Call).

Lucy began to lay at almost five months, and Whoopee and Zsa Zsa kicked in at five and a half months. Lucy is the most prolific chicken, with almost 300 eggs in her first year of laying. Zsa Zsa had laid 200 eggs in almost eleven months. Whoopee is the laggard, tallying a mere 112 eggs over ten months. That's okay — what Whoopee lacks in egg production, she makes up for with her good looks and goofy personality.

How do I know how many eggs the Girls have laid? Because I keep an egg journal. It's a fun way to keep track of the eggs I collect from my flock. You can use a calendar, a spreadsheet, or even a real journal to document your hens' egg-laying ways. Once the Girls started gifting me with eggs, I became familiar with the color and style of egg laid by each chicken. I identified and noted each egg every day on my kitchen wall calendar. Of course, keeping track of eggs is easy with two or three chickens, but it gets more challenging with each additional hen.

## WHEN HENS START LAYING

Hens begin laying anywhere from five to six months of age. A pullet's first egg will be small. It may also be discolored and somewhat misshapen. That's okay — after all, your chicken is just getting started, and it takes a couple of months for nature to calibrate this process. After that first egg, your hen may not lay for another day or two. When she does lay again, the egg will be larger and the color more firmly established.

If you pay attention to your hens, their behavior will let you know when they are getting ready to start laying eggs. The first is what I call the Egg March. Although the hens don't know it, nature is getting them ready to lay. The Girls were about four months old when I noticed that at about the same time each day, all three hens, if loose in the yard or just hanging out in the coop, would resolutely march back into their henhouse. I peeked in on them to see what they were doing. They were standing around, taking turns digging in the nest box. Half an hour later, they'd go back into the yard, unsure why they'd left it in the first place.

The other thing a hen will do when she is about ready to lay (or when she's just started laying) is squat down in a defensive posture, wings slightly away from her body, when you try to pet her. At first I thought my hens had developed a severe inferiority complex. Then I learned that this is the position a breeder takes when she's getting ready to accept a mate. Mating, of course, is coincident with the production of eggs.

When a hen lays her first egg, she's not sure what is happening to her. The urge to lay comes on suddenly. She will retreat to the nearest corner of the yard or coop. There, she will spend the next 15 minutes diligently digging a shallow hole. Once the hole is a depth acceptable to the hen, she will sit in it, tail to the wall, beak to the wind. Another 15 minutes go by, and she suddenly stands up and runs off as if nothing has happened. There sits your first egg, snuggled in a nest of fallen leaves and still hot to the touch!

If your hen starts laying in mid-summer, she will be pretty regular, with an egg every day or two. If a hen begins to lay in fall, or in winter after the fall molting, eggs won't come so regularly — maybe two to four eggs each week. Hens lay better in warmer weather, and they slow down in the cold season. However, if the weather gets too warm and the hens can't cool properly, their laying may be less consistent.

Chickens lay their eggs at about the same time each day. Because chickens typically lay an egg every

## NEST BOX NEWS

■ ■ ■ ■ ■

Keep an egg journal
to track your chickens'
egg-laying prowess.

■ ■ ■ ■

Hens lay their eggs in
"nest boxes," private,
shoebox-sized cubicles
along a henhouse wall.

25 hours, you would expect them to lay their next egg later each day. My Girls don't do this. Lucy is the early riser, and I'll find her hot egg between 8 and 10 A.M. each day. Zsa Zsa jumps into the nest box next and starts her celebratory egg squawk about an hour after Lucy. True to her laid-back form, when Whoopee bothers to lay an egg, it's late in the afternoon, after I've already gathered the early eggs. Though Lucy and Zsa Zsa always lay eggs in their nest boxes, Whoopee sometimes can't be troubled to waddle into the henhouse, and she will drop her egg just inside the coop door, where she was pacing and waiting to be let out.

## HOW TO GET EGGS *IN* THE NEST BOX

Encourage your hens to lay in the nest boxes you've built for them. To do this, keep your hens cooped up until late in the afternoon, until you're sure they all had a chance to lay. This way, the hens won't get used to laying hidden eggs in the garden. Once your hens are accustomed to their laying routine, they can be let out in the yard, and when the urge to lay hits them, they'll run right back to the nest box.

Sometimes a hen just won't get it. Instead of laying an egg in those lovely nest boxes you've built her, she will lay on the henhouse floor (not the cleanest place to lay eggs, especially after a night of roosting) or dig a hole in the run and lay there. Show your hen to the nest box by planting a fake egg there. You can buy a fake egg at the feed store, but any egglike object will do. I've used those hollow plastic Easter eggs and golf balls. Both work well in luring the hen to the nest box to lay. Once a hen sees an "egg" in the nest box, she seems to say to herself, "Ooh, that looks like a good place to lay an egg! Someone has already laid there . . . guess I will lay there, too!" Voilà, egg in nest box.

## WINTER LAYING

Chickens need between 14 and 16 hours of light each day to lay. To keep your hens laying eggs consistently through the winter months, install a hanging light fixture with a 25- to 40-watt bulb set on a timer in the henhouse. Some folks install heat lamps for this purpose, but I think that they run the risk of making the henhouse too hot. Set the timer to turn on two hours before dawn and two hours after sunset. Small-wattage bulbs in the henhouse can make a big difference in the frequency and number of eggs laid throughout the winter months.

While artificial winter lighting for chickens is generally a good thing, I have one caveat. Extending natural daylight appears to play around with chickens' internal clocks. They wake up before the sun comes up, and they stay awake in their henhouse long after the sun has set. The net effect can be to turn mere household hens into bona fide party chicks.

I gave the Girls lots of extra light last winter. I noticed that instead of going to bed at dusk, like most chickens, the Girls began hanging out later and later in the garden. I wasn't sure how they were doing this, as chickens are notoriously night blind. Then I saw that they were using the shaft of light that shone through the open coop doors to find their way back home. One time, the coop and henhouse doors had somehow closed, and no light spilled out into the yard. On a moonless winter night, I looked out the back window and saw them standing together, fluffed out and huddled close in a group in the middle of the lawn. They had been so busy hunting bugs they didn't notice it had become dark. The Girls couldn't find their way back to the henhouse and figured they'd tough the night out unsheltered in a huddle in the yard.

## COLLECTING THE EGGS

In an ideal world, you would collect eggs promptly after they are laid. But because most folks aren't stay-at-home chicken lovers, it's more realistic to say that the eggs should be brought in no later than the end of each day. Eggs left in the nest box overnight or for several days will not be fresh. Also, eggs left too long in a nest box are prone to breakage, as the hens lay more eggs on top of preexisting eggs. Leaving eggs in the nest box may encourage bored or anxious hens to begin eating the eggs. This bad chicken habit is very hard to break once it's begun, so prevent it by collecting eggs each day. This won't be too hard to do, especially if you have kids. Kids love to race outside and get the fresh eggs. Who can blame them? Collecting fresh eggs from your own garden is really cool!

Bits of dirt or manure may adhere to the eggshell during the egg's brief stay in the nest box. When you bring in the eggs, wipe them clean with a dry cloth or lightly scrub them with a coarse paper towel or very fine sandpaper. If you can avoid it, do not wash the egg with water. Eggshells have a natural outer coating that keeps bacteria out. This outer coating is water soluble; in other words, if you wash the egg, you also wash away its protective coating. If the egg is so dirty that you must wash it, use it right away.

If you can't use a dirty egg right away, use a damp sponge with no detergent to rub off the dirt. Dry the eggshell thoroughly with a soft cloth, then rub a little cooking oil over the entire egg, wiping away the excess. The oil will help replace the natural coating on the shell that keeps bacteria and other unwanted organisms out of the shell's precious contents. Personally, I don't fuss around much with Ma Nature, and I recommend washing the egg only as a last resort. Usually a dry rub will do the job.

Store eggs in the refrigerator, where they'll stay fresh for two to three weeks.

Do not hard boil eggs immediately upon collecting them. The whites and yolks of day-fresh eggs just don't gel adequately. Also, the egg white sticks in large chunks to the eggshell when you try to peel a fresh hard-cooked egg. I recommend that, after collecting eggs, you dry them and put them in the fridge for a day or two to give their gelatinous contents a little time to settle down inside their shells.

**BIRD WORD**

A chicken can lay more than 600 eggs in her first two years.

▪ ▪ ▪ ▪

If an egg you collect is dirty, wash and dry it, then rub it with cooking oil.

Although I doubt you'll have trouble giving away your fresh eggs, you just might have Super Hens that lay too many. Don't let any precious eggs go to waste. You can always freeze eggs for later use. However, never freeze eggs in their shell unless you enjoy eggs exploding in your freezer. Crack the fresh, surplus eggs into a freezer-tolerant container, scramble them up with a little salt, and seal the container securely. This way, your extra eggs keep in the freezer up to six months.

If you and your family are eating all the eggs you want but

don't want to freeze the extras, you may want to share the surplus with those less fortunate than you and your family. Contact your nearest women's or homeless shelter or any other charity that feeds hungry people to inquire whether it will receive your eggs.

What if you have more eggs than you can give away? Can you sell your surplus? The only thing you cannot readily do with your surplus eggs is sell them. Commercial egg producers are required to heed federal standards and industry guidelines for egg quality, so all eggs (presumably) are inspected to ensure they comply. Unless you want to set yourself up as a commercial egg producer,

*Poultry Tribune*, circa 1940.

complete with adherence to federal regulations, regular health inspections, and a lot of other red tape, you cannot sell your eggs. In fact, city law usually prohibits the unregulated sale of fresh eggs by residential chicken keepers. Exceptions to this type of law may permit you to sell the eggs privately at produce stands and community farmers' markets. However, don't sell a single egg until you check your city codes or town ordinances for the law.

# Coop Care

Many people associate chickens with filthy living conditions. It's true that a chicken coop that's not cleaned once or twice a week will get stinky. But we human keepers have certain responsibilities to our critters. The main commitment that we have to city chickens is to keep their coop clean. If the coop is well maintained, it won't smell, and the birds will be both happy and healthy.

Large numbers of chickens are difficult to keep super clean. That's why chicken farms are located on acreage, not on a city lot.

Fortunately, a few chickens in a city yard are nothing like dozens of chickens on a small farm or hundreds of birds in a mass-production chicken environment. The advantage to a small flock of chickens in the city is that the birds are easy to keep clean. You don't need a lot of fancy equipment, regular inspections from the USDA, or even more than a couple of hours of work every week. Cleanliness for a small two- to five-bird coop simply means removing and properly disposing of the manure in the coop and henhouse regularly.

Moisture is the enemy of a clean coop. Excessive moisture from chicken droppings or spilled drinking water leads to molds and bacteria, which can lead to allergies or infections in the chickens. Moisture unchecked eventually ferments and stinks, which leads to unhappy neighbors and a steady convention of blowflies in and around the coop.

Your chicken coop will stay fresh and dry if you place liberal portions of chopped straw and cedar shavings on the floor of the run and the henhouse. It's important that the straw be chopped; unchopped straw can contribute to moisture buildup and disease. Each week, spread 8 to 12 inches (20–31 cm) of chopped straw over the chicken run. It sounds like a lot, but it gets tamped down quickly with a busy flock of bell-bottomed chickens stomping about. Spread 1 to 2 inches (3–5 cm) of cedar shavings over the floor of the henhouse, and put fresh chopped straw over that, in the same thickness as in the run. Cedar shavings are sweet smelling and absorbent; they'll provide an extra measure of protection against moisture and odor. Fill each nest box with 6 to 8 inches (15–20 cm) of straw, tamping it down to provide a firm fit for the hen and her bottom. Hens like to lay their eggs in tidy places that are private and quiet.

You can buy bales of straw from your local feed store. Large bales of straw cost about $5 and will last for several coop cleanings. They should be stored in a dry place. I buy two bales at a time and store them behind the chicken coop, tucked beneath the deep roof eaves and covered by a rubber tarp. Don't store your straw where the chickens have access to it. Chickens love to play in fresh straw. If you leave a bale where they can get to it, they will have the time of their lives kicking it apart.

At least once a week, remove thickly soiled sections of straw from the run and henhouse. Scoop up the debris with a pitchfork or rake it into a pile. If you don't have a compost bin, now is the time to get one. Put the soiled straw into your compost bin, along with the usual organic composting materials. Chicken poop is a rich source of nitrogen, an essential active agent in the composting process. After you have scooped out most of the droppings in the coop, fluff up the remaining straw with a pitchfork or rake. Then, top it off generously with a layer of fresh straw 8 to 12 inches (20–31 cm) thick.

If you don't have enough space for more than one or two compost bins, you need to consider how and where to dispose of soiled straw. On trash collection day, you could put the soiled straw out in addition to other yard debris for the period. A better alternative is to offer it to your friends. I have friends and neighbors lining up for my manure-laced straw. They bring by their trash cans and wheelbarrows, and I fill them up with the soiled bedding from the coop. Friends leave my house with cars full of chicken poop and smiles on their faces. I'm never at a lack for takers on coop-cleaning day. Actually, I have a waiting list for every barrel of poop straw I don't use in my own compost bin. Who would have guessed that chicken poop would make me so popular with my friends?

**Chick Chat**

Straw is the best kind of litter for the chicken coop; it is absorbent and sweet smelling.

∎ ∎ ∎ ∎

Composted chicken manure becomes rich garden soil.

A note about composting: It takes time. Warmth, combined with regular air and moisture, hastens the composting process, regardless of what is in the compost bin. Don't expect to toss mounds of chicken droppings into the compost bin and then to mix it into your garden soil a week later. The high acidity of the chicken manure in the immature compost can be too much for your garden plants. Putting large, concentrated quantities of chicken manure directly into a garden can "burn" your plants. Manure, together with the other compost ingredients (lawn clippings, fallen leaves, coffee grounds, vegetable discards — produce too wilted or rotten to feed to the chickens — and bits of clay soil) needs to break down over three to six months.

After a few weeks, the dirt in the chicken run, if once compacted clay soil, will be dark and crumbly. The chicken droppings and straw, together with the nearly constant scratching by the hens in their pen, instantly improve any soil beneath their scaly chicken feet.

## THE DEEP LITTER SYSTEM

Some small-flock chicken keepers have another method of dealing with coop care — they do almost nothing. This is known as the "deep litter system," and it works like this: no removal of droppings, no fluffing, just adding straw regularly to the coop and henhouse floors all summer through winter. In spring, all of that prime compost material has been mulching nonstop for three seasons. This is one way to efficiently compost chicken waste and organic matter directly onto a large area without doing much of anything. However, you have to control the odor that will surely emanate from the coop by leaving the chicken waste in and merely adding lots of straw and cedar shavings. The problem with the deep litter system, besides the inherent odor worry, is the tendency for the coop to get damp. Dampness is not a condition tolerated by chickens for long without some type of disease sprouting up in the flock (like coccidiosis, a parasitic infection transmitted through chicken droppings). Personally, I prefer the meager labor once weekly in freshening up the chicken coop. The deep litter system would work better in a small, mobile coop or in a coop on a large tract of land with not many neighbors nearby. For urban chickens, the deep litter system simply stinks too much.

## MAINTENANCE

Every couple of months, inspect the coop and henhouse to make sure there are no loose screws or nails, splinters, or jagged wire that could harm your flock. Chickens love to peck and pick at things with no regard for their own personal safety. They'd swallow a small screw in a jiffy if given a chance, so don't give them the opportunity. Experience has taught me that important lesson.

I have a basement window that looks into the coop. I had noticed that the caulking was missing from the frames; panes were loose. This could be dangerous for the Girls if one of them pushed up against the glass. I put up the temporary garden fence and routed the hens out of the coop. Then I caulked the panes with generous helpings of wood putty. Hours later, the putty appeared dry and I bribed the hens back into the coop with — what else — cottage cheese. Early the next morning, I woke to a distant, rhythmic tap-tap-tapping. It sounded like it was coming from inside the house.

## NEST BOX NEWS
■ ■ ■ ■ ■

Every couple of months, inspect the coop to make sure it has no loose screws or nails, splinters, or jagged wire that could harm your flock.

Thinking rats were rummaging around in the basement, I quietly went downstairs to sneak up on them. The tapping noise grew the lower I got. Tap-tap-tap. I looked around the dark basement. It took me a minute, but I realized that the hens were making the tap-tap-tap as they feasted on the window putty from inside their coop. Oh no! I ran outside to the coop and shooed the Girls away from the window putty. Fortunately, it was just one hen (Zsa Zsa) who did most of the eating — it was easy to tell because she was the only hen with a full putty "beard" outlining her beak. Fearing the worst but hoping for the best, I promptly fed her an apple, hoping to flush out the putty. I was lucky. Zsa Zsa seemed to have no ill effects from consuming several tablespoons of wood putty, and I didn't lose a hen. I reputtied the pane, this time using a super-fast-drying brand. I also put up a wire screen around the window so the Girls couldn't get right up next to it. No more putty for the poultry.

## CHORE CALENDAR

The scoop on coop care is basic and unwavering: Keep the coop clean all the time. No matter when you do it, or how you do it, do it. Your chickens and your neighbors will appreciate the fresh straw and clean accommodations. Remember, if the coop smells, it's not the chickens' fault . . . it's yours!

**Daily**
- Check water; refill if low; remove debris from drinking tray
- Check food; refill if low; remove debris from eating tray
- Collect eggs

**Weekly**
- Clean run and henhouse: remove manure droppings and dirty straw; add clean cedar shavings and straw
- Wash water dispenser
- Check grit dish; refill if low
- Check for rodent holes into run or henhouse

**Monthly**
- Visit feed store for straw, chicken feed, grit, and scratch
- Call friends to pick up straw for their compost bins

**Yearly**
- Make repairs to run and henhouse, if necessary

# Extreme Weather Care

In summer, keep an eye on your chickens to make sure they are not too hot. Provide them with plenty of water. A heavy breed of hen can drink more than a quart of water each day during warm weather. Make ample shade available during the hottest part of the day. The henhouse can become an oven when the thermostat hits 90° F. Don't let your chickens roast on the roost! Ensure the henhouse is adequately ventilated. Open its doors and windows to let any stray breezes drift through. If the day is hot and still, install a small portable fan in the coop by attaching it to the plug outlet used by the henhouse heating lamp in winter.

When chickens get too hot for their comfort, they pant. Their beaks part and stay open, and their little chicken tongues stick out with each labored breath. When the Girls look really pitiful, I let them out into the garden. They dig shallow holes in shaded soil and lie, breast down, wings spread open on the ground, until they cool off. If they are still hot, I fill a clean spray bottle with cool water and mist them.

Bantams aren't as susceptible to overheating as their heavy breed cousins, but they are sensitive to freezing cold temperatures. When keeping bantams in your flock, be sure to keep them in an enclosed, draft-free, heated henhouse during any cold spells.

What's the optimal temperature for a henhouse in winter? Depends on how cold it is and what kind of chickens you have. Heavy breeds, like the Girls, need a heat lamp when temperatures are at or below the freezing point. Bantam breeds need a heat lamp when outside temperatures dip into the 40s.

Use a 25- to 40-watt light bulb or the heat lamp from the chicks' brooder to warm the henhouse. Affix the light or lamp away from straw and not directly over the hens' perch. The heat from these bulbs will raise the temperature in a 3' x 5' x 5' (0.9 x 1.5 x 1.5 m) henhouse ten degrees over the course of several hours. Put the light or lamp on a timer so that it comes on around midnight and turns off when the sun comes up.

Do not use an oil or electric heater to warm the henhouse. Such devices provide too much heat for the compact space of a henhouse. The last thing you want is to dehydrate your poor chickens!

# Health & Fitness

Four key elements make up a chicken health care program: a clean environment, plenty of fresh food and water, protection from the elements, and exercise.

Careful attention to coop care will ensure that your chickens have a clean environment. Nothing makes chickens (or any other pet, for that matter) sick quicker than filth.

Healthy hens are hungry hens. Eating is one of a chicken's main hobbies and reasons for living, so don't deprive your hens of munchies. Make sure the hens' feed is fresh and plentiful.

Healthy chickens are dry and warm chickens. Make certain the henhouse is watertight. It's difficult to make the coop completely water-free, so give the chickens one dry place to go in their quarters. A damp chicken is prone to catching colds (yes, chickens can catch colds!) or developing infections.

If you notice that your chicken has a cold (she will be sneezing and sniffling, same as you), crush some fresh garlic and mix it into her scratch, or put about a teaspoon of fine garlic powder into a gallon of the hens' drinking water. The curative marvels of garlic do not discriminate between people and chickens, and soon your flock will feel better.

Healthy chickens need their exercise. They love to walk and scamper around the garden, scratch up dirt for bugs and worms, and run and flap their wings a bit (this latter activity may look like your chicken is trying to fly, but she won't make any progress, especially if your chicken is a plump, heavy breed). This outdoor time breaks up the boredom of staying in the coop all the time. If your hens don't get enough leg time in your garden, it can cause them to become anxious and fidgety, and they may start picking on each other. Regular free-range grubbing and running in the yard make for happy, well-adjusted chickens.

If I don't let my hens out for a couple days because of bad weather, they'll get cranky and stop laying for a few days (okay, Girls, I get the message). But as soon as they get some yard time, even half an hour between rainstorms, they start laying eggs again. It just goes to show that regular exercise is as important to a chicken's health as it is to ours.

## EGG BINDING

Chickens that don't get enough activity get fat. Obesity in chickens contributes to the health problem known as egg binding, which occurs when an egg has become lodged up inside the hen, and she cannot expel it during laying. A fatty layer has built up around the hen's reproductive organs, inhibiting the oviduct from moving the egg down toward the vent. Sometimes chickens bind up simply because an egg they are trying to lay is too large.

If you notice that your hen is looking uncomfortable, standing or moving in a strange way with her head or tail down, she might be binding up. Help her immediately! Place her in a warm area (under a heat lamp) to relax her. Sometimes warmth alone will help her pass the egg. If heat doesn't work, massage some vegetable oil very carefully around her vent while gently massaging the hen's stomach in the direction the egg should be traveling. If no egg emerges, hold the hen over steaming water, being careful not to burn her. I've read that some folks, as a last resort, will gently break the egg while it is in the hen, and bit by bit empty its contents, with the hen's contractions pushing out the rest. However, the hen may die from shock or a heart attack if the egg remains stuck in or while the egg is being broken inside her. Exercise your hens consistently to avoid this and most other health problems.

## FEATHER PICKING

Another health problem for chickens is feather picking, which occurs when your chicken, usually prompted by some type of physical or emotional discomfort, pulls off large quantities of its own feathers or the feathers of fellow hens. If you catch your chicken at feather picking, you can discourage this behavior by applying a mix of one part vinegar to five parts water with a washcloth to the bird's bare areas. This cleanses the area, and the stink and taste of the vinegar discourage the chickens from further picking.

## MITES

Chickens take dust baths to cleanse off any mites they pick up. However, if the mites are visible to your naked eye, the problem is too far gone for the hen to take care of with a dirt bath alone. Be on the lookout for a mite infestation: If one chicken has mites, chances are the others in the flock also do. Mites sometimes appear as tiny, almost invisible brown pinheads crawling on a chicken's skin, beneath the feathers. The mites may also be visible crawling on the roost where the chickens sleep. Another type of mite demonstrates itself as patchy, crusty, whitish scales on a chicken's legs.

You can eliminate mites with either a powder or a spray treatment available at the feed store or on-line hatchery and supply stores. These powders and sprays aren't toxic and are usually effective with repeated applications. To apply the powder, place the recommended dosage in an area in the coop where the chickens take their dust baths. This way, they can apply the dust themselves without being handled. If you apply the powder or spray directly on the chickens, I recommend doing this at night after they have gone to roost. Chickens don't see well in the dark and are quiet and passive after sundown. You can control exactly where you apply the medication, and the chicken won't stress out and protest the application as they would during daylight hours.

A home remedy for leg mites is to soak the chicken's legs in warm, soapy water twice daily until effective. However, holding a chicken still long enough to really soak those legs is more of a challenge than some could handle. Instead, use a damp, soapy sponge or washcloth to wipe the chicken's legs with one hand while holding her under your other arm. Again, to make this treatment easier for you and the chicken, do it after sundown.

## Molting

Like any other bird, your chickens will molt each year. Chickens start molting as early as midsummer, and the molt may continue until late fall. During their molt, chickens lose old feathers and new feathers, called *pinfeathers*, sprout out like porcupine spines. Some chickens molt so slowly and gradually that you barely notice they're molting. Others throw off all their feathers at once and are half-naked for a couple of months.

Molting is stressful for a chicken. Just imagine if you spent your entire life covered in soft feathers, then suddenly lost them all and became covered in hard-shelled spikes instead. You'd probably be uncomfortable, too! Chickens aren't real happy when they molt. In fact, they get downright crabby. Sometimes they become skittish, moody, and aggressive. And they lay considerably fewer eggs.

Make sure your chickens get plenty of good, nutritious feed during the molt. This will help them maintain their strength and vigor. A vigorless hen is a sorrowful sight.

■ ■ ■ ■ ■ ■ ■ ■

For more information on chicken health and home care for certain ailments and conditions, check out *The Chicken Health Handbook,* by Gail Damerow. It has everything you'll ever need to know about chicken health through the years of keeping your pet flock. If you are not comfortable administering health care other than the basics, or if your chickens are too sick for home remedies, contact your local avian medical center or veterinarian specializing in bird medicine and surgery.

CHAPTER

8

# Eggcellent Eats

A FRESH EGG STANDS OUT in color, flavor, and consistency. The yolks are a dark, rich orange-yellow. That's because urban chickens, unlike their commercial cousins, get lots of greens and other produce containing beta-carotene, which gives fresh yolks that brilliant orange hue. Both yolk and white stand firm in a pan. The flavor is rich and robust.

My first fresh garden-grown egg was fried in a dab of butter and seasoned with salt and freshly ground pepper. The yolk, when gently forked, casually oozed out like sweet orange lava. It was incredible. The next fresh egg to grace my plate was soft boiled. I ate it in the same way I did when I was a kid, with a little salt and pepper in each bite, dipping toasted, buttery bread into the warm yolk.

With a continual bounty on hand, I've been amenable to tasting my fresh eggs with something other than salt, pepper, and butter. Here are a few of my favorite egg recipes, all of them real simple to allow the tasteful splendor of fresh eggs to shine through.

# Egg Sandwich

1 tablespoon butter
1 egg
2 slices bread (your favorite)
1 slice Cheddar or Swiss cheese
1 sausage patty or strip of bacon, cooked
Salt and freshly ground black pepper

In a nonstick frying pan, melt the butter on medium heat. Crack the egg into the pan. Toast the bread. Put the cheese on one slice of toast, then cover with the other slice to gently melt the cheese. When the egg is cooked the way you like, slide it onto the slice of bread with the cheese on it. Top with the sausage patty, add salt and pepper to taste, and cover with the remaining toast slice. Cut the sandwich in half to serve.

# Quesadilla Scramble

4 eggs
½ cup milk
Salt and freshly ground black pepper
2 tablespoons butter
¼ cup chopped onion
1 cup cooked rice, warm
4 corn tortillas
1 cup shredded Cheddar or Monterey Jack cheese, shredded
4 slices bacon or 4 sausage links, cooked and cut into 1-inch pieces
Salsa
Sour cream
Pickled jalapeños, sliced
Cilantro, washed, dried, and chopped

In a small bowl, beat the eggs with the milk and add salt and pepper to taste. In a nonstick frying pan, melt 1 tablespoon of the butter on medium heat. Add the onion and cook 3 minutes. Add the egg mixture to the pan, and scramble the eggs. Remove the eggs from the heat and transfer them to a bowl until ready to assemble the quesadilla.

Melt the remaining butter in the frying pan. Add one tortilla at a time, cooking both sides of each until golden (about 1 minute per side). Drain on paper towels. Place one tortilla on each of four dinner plates. Top with ¼ cup of the cheese, then rice, one-quarter of the scrambled eggs, the bacon or sausage, and the condiments.

# Egg Burritos

4 eggs
½–¾ cup milk
Salt and freshly
    ground black pepper
8 slices bacon, cut into
    1-inch pieces
1 tablespoon butter
½ cup chopped onion
¼ cup chopped
    jalapeños
4 flour tortillas (10
    inches or larger)
½ cup shredded Cheddar
    or Monterey Jack
    cheese
Salsa or hot sauce
Sour cream
Guacamole
Cilantro, washed,
    dried, and chopped

In a small bowl, beat the eggs with the milk (more milk for fluffy eggs, less for denser eggs) and add salt and pepper to taste. In a nonstick frying pan, cook the bacon on medium heat to your liking. Drain the bacon on a paper towel. Discard the grease and wipe the pan. Add the butter to the pan and melt on medium-high heat. Add the onions and cook about 3 minutes. Add the jalapeños and cook another 3 minutes, stirring the mixture to prevent sticking. Add the egg mixture and lower the heat to medium. As the eggs cook, use a spatula or wooden spoon to raise the cooked areas of eggs, and tilt the pan to direct uncooked eggs to the raised area. If you want fluffy eggs, don't disturb the cooking eggs too much. If you like them scrambled dry, mix thoroughly. Cook the eggs until they're done and remove them from the heat.

Put a flour tortilla on a plate. Sprinkle one-fourth of the grated cheese in its center. Immediately cover the cheese with one-fourth of the cooked eggs. Add the bacon and condiments. Roll the burrito closed and serve. Repeat with the remaining tortillas.

# Eggs Benedict

4 eggs
2 English muffins, split
4 slices ham or Canadian
   bacon
   Hollandaise Sauce
   (see recipe below)
   Paprika or dill

In a large, shallow skillet, bring 1 inch of water to a boil. Reduce to a gentle simmer. Crack one egg into a shallow glass dish. Begin toasting half of a muffin. Gently slide the egg into the simmering water. Heat the ham in a frying pan while the egg is poaching, 2 to 3 minutes.

Remove the muffin from the toaster and set it on a plate. Place the hot ham on the muffin. Use a slotted spoon to remove the egg from the skillet, and set the egg on the ham. Top with a generous ladle of hollandaise sauce and sprinkle with paprika or dill. Repeat for remaining servings.

# Hollandaise Sauce

3 egg yolks
2–3 tablespoons lemon juice
   Salt and freshly ground
   black pepper
$\frac{1}{2}$ cup unsalted butter, melted
1–2 tablespoons water
   Hot sauce

In the top half of a double boiler (or a stainless-steel bowl that sits in a pot of water, water not touching the bottom of the bowl), whisk the egg yolks with the lemon juice and salt and pepper to taste, until the mixture is pale yellow. Remove from the heat and whisk in the melted butter, a little at a time, until thoroughly blended and smooth. Add water to achieve the desired thickness and hot sauce to taste. Keep on very low heat and stir occasionally until needed.

# Deviled Eggs

6 eggs (at least 1 week old)
¼ cup mayonnaise
⅓ cup sour cream
1 tablespoon relish
1 teaspoon Dijon mustard
1 teaspoon horseradish
1–2 dashes hot sauce
Salt and freshly ground black pepper to taste

FOR THE GARNISH
Paprika or cayenne pepper

Place the whole eggs in a 2-quart pan and cover with cold water. Heat on high until boiling, then reduce the heat to medium-high and cook for 10 minutes. Remove from the heat and drain off the hot water. Rinse the eggs with cold water several times, and fill the pan with cold water to cool them. When they have cooled, peel off the shells.

In a medium-sized bowl, mix the remaining ingredients. Cut the eggs in half lengthwise. Pop the yolks into the bowl. Place the egg white halves on a plate, hole side up. Mash the yolks and mix them thoroughly with the mayonnaise mixture. Add more mayonnaise or sour cream for a creamier mixture. Using a teaspoon or pastry bag, fill the egg white halves with the yolk mixture. Sprinkle each egg with a dash of paprika or cayenne. Serve chilled.

# Caesar Salad

1 egg
3 garlic cloves, finely
   chopped
2 tablespoons Dijon mustard
2 tablespoons lemon juice
½ cup extra-virgin olive oil
   Anchovies (half of a small
   can, drained)
   Salt and freshly ground
   black pepper
1 head romaine lettuce,
   washed, dried, and torn
   into pieces
½ cup grated fresh Parmesan
   or Romano cheese
   Croutons (see recipe
   below)

Coddle the egg (boil it for 30 to 60 seconds).

In a food processor or blender, combine the garlic, mustard, lemon juice, olive oil, and anchovies and chop fine. Scrape the sides of the bowl and add the coddled egg and salt and pepper to taste. Blend until smooth. Pour half of the dressing into a large bowl. Add the lettuce and toss. Add additional dressing as needed to coat the lettuce. Add the cheese and croutons. Toss thoroughly and serve.

# Croutons

1 tablespoon extra-virgin
   olive oil
   Oregano and basil
2 slices bread (any kind),
   cut into 1-inch pieces
   Salt

Heat the oven to 350° F. Heat the oil on medium heat in a nonstick frying pan. Add oregano and basil to taste. When the oil is hot, add the bread squares. Season the tops of the bread with salt to taste. Turn to cook both sides until golden brown. Put the toasted cubes onto a cookie sheet, and place in the oven for 10 minutes, or until the croutons are crispy.

# Quiche with Sausage and Asparagus

**FOR THE CRUST**

| | |
|---|---|
| 2 | cups all-purpose flour |
| 1/4 | teaspoon salt |
| 1 | cup unsalted butter, cut into 1-inch chunks |
| 2–3 | tablespoons water |

**FOR THE FILLING**

| | |
|---|---|
| 4 | eggs |
| 1 | cup milk |
| 1/4 | cup half-and-half |
| | Dill |
| | Salt and black pepper |
| 1 | cup grated Swiss cheese |
| 1 | cup cooked asparagus |
| 8 | ounces Italian sausage, cooked and drained |

TO PREPARE THE CRUST: Heat the oven to 350° F. Combine the flour, salt, and butter in the bowl of an electric mixer with the paddle beater. Beat on low for about 4 minutes, until crumbly. Add the water to form the dough. Press the dough into a 10- to 12-inch quiche pan with a removable bottom (or, in a pinch, a pie pan will do). Bake 12 minutes, remove from the oven, and set aside to cool.

TO PREPARE THE FILLING: Heat the oven to 350° F. Whisk together the eggs, milk, and half-and-half. Add the dill, salt, and pepper to taste. Spread the cheese on the bottom of the cooled crust, then top with the asparagus and sausage. Pour the egg mixture over the crust contents. Bake 30 minutes or until the quiche appears set. Cool for 10 minutes and serve.

# Pound Cake

1 cup unsalted butter, plus extra for greasing the pan

2 cups all-purpose flour, sifted, plus extra for sprinkling

1¼ cups sugar

1 teaspoon vanilla extract

4 eggs

½ teaspoon baking soda

½ teaspoon salt

Heat the oven to 325° F. Butter a 9- by 5-inch loaf pan, then coat the pan with flour, shaking out any excess. Make sure the eggs are at room temperature. If the eggs are cold, cover them with warm water for 3 minutes.

Place the butter in a large mixing bowl; beat until creamy. Add the sugar a quarter cup at a time while beating with an electric mixer on medium until thoroughly blended. Mix in the vanilla. Add the eggs, one at a time, to the butter mixture and blend well after each addition. Stir in the flour, baking soda, and salt. Pour the batter into the loaf pan and spread evenly. Bake 1 hour or until a wooden skewer comes out clean. Cool 5 minutes in the pan, then on a wire rack.

VARIATIONS·

- Chocolate chips — add a couple of handfuls.
- Lemon — add 1 tablespoon lemon juice and ¼ teaspoon lemon zest.
- Orange — add 1 tablespoon orange juice and ¼ teaspoon orange zest.
- Buttermilk-pecan — add 1 cup buttermilk, 1 cup flour, and ¼ cup chopped pecans.
- Whiskey-walnut — add ¼ cup whiskey, ¼ cup flour, ¼ cup walnuts, and ¼ cup raisins.

# Flourless Chocolate Raspberry Cake

1 cup unsalted butter, plus extra for greasing the pan

8 ounces semisweet chocolate squares, broken in half

2 teaspoons vanilla extract, Kahlua, or whiskey

8 eggs, separated

¾ cup sugar

½ cup raspberry jam
Chocolate Ganache (see next page)

Heat the oven to 325° F. Liberally grease a 9-inch springform pan with butter. In the top half of a double boiler (or a stainless-steel bowl that sits in a pot of water, water not touching the bottom of the bowl), melt the butter and chocolate, about 10 minutes. Remove from the heat and add the vanilla.

With an electric mixer, beat the 8 egg yolks with the sugar on high speed until the mixture is pale yellow. In a separate bowl, beat 4 of the egg whites on high speed until firm peaks form. Discard the remaining four whites. Fold the chocolate mixture into the yolk mixture. Add one-third of the egg whites, incorporating them thoroughly. Fold in the remaining egg whites. Pour the batter into the pan and bake for about 75 minutes, or until a toothpick inserted in the center comes out clean. Cool in the pan for 10 minutes, then invert the cake onto a plate. Refrigerate at least 1 hour.

Top the chilled cake with a thin, even layer of raspberry jam. Refrigerate for another hour. When the jam is chilled on the cake, pour the warm ganache over the top and sides of the cake. Serve immediately, or after chilling for 15 minutes.

# Chocolate Ganache

6 ounces semisweet
   chocolate squares,
   broken in half
2 tablespoons unsalted
   butter
2 tablespoons corn syrup
$\frac{1}{4}$ cup whipping cream
1 teaspoon vanilla extract,
   Kahlua, or whiskey

Melt the chocolate and butter in the top half of a double boiler, about 8 minutes. Remove from the heat and blend thoroughly. Add the corn syrup, whipping cream, and vanilla and stir until smooth. While the mixture is still warm, spread it over the cake.

# Epilogue

■■■■■

## A Day in the Life
## of an Urban Chicken Keeper

P URPLE CLOUDS STREAK ACROSS THE EASTERN SKY as the sun rises behind Mount Hood. Below the dozing volcano, a city erupts from sleep to wakefulness in a startled fit. Automatic coffee brewers click into action; electric razors slide across thousands of chins; school bus brakes squeal in protest at every single stop. It is morning, and the city of Portland is waking up.

The Girls and I are still asleep. We won't awaken for another hour (lucky us). Finally, my automatic coffee pot ticks on. When I hear the last garble of steam, I arise. The Girls don't plunk down from their perch until they hear me clank around the kitchen, trying to pour coffee in my cup and not on my toes.

I take the cup and go into the laundry room. I can see the back-yard and into the chicken coop from the side window. Sipping hot coffee, I absently watch the Girls. For chickens, they are late risers and slow starters. They're standing in the coop now, lazily preening, getting ready for the day. Whoopee hears me set my cup on the windowsill and stops preening. A beady eye turns up and locks with my eyes. A dim recognition goes off in her bird brain, and she realizes it is me. She steps over to the coop door and pushes against it with her chicken breast. The door is locked. It always is. She always tries.

Whoopee starts to pace by the coop door. The other two Girls have discontinued their morning ablutions and joined her. I look down at my flock and smile. They look up at me and hope I'll feed them. They seem not to notice the giant bin full of chicken feed hanging behind them. They keep looking up, six purportedly starving eyes. I leave the window and head into the kitchen to make my breakfast. If the Girls knew, they would be furious.

As I'm working at my desk, a disgruntled clucking cracks the mid-morning quiet. I turn on Chick TV to see what is going on. Camera 1: Whoopee still by the coop door, scratching in the dirt and kicking straw into the watering tray. Camera 2: No subjects. Camera 3: The source of all that clucking.

Lucy, the early layer, is trying to sit quietly in a nest box. Zsa Zsa harasses Lucy by standing on the lip of the adjacent nest box, hovering in Lucy's personal space. Despite having access to three cozy nest boxes, the Girls all favor the box nearest to the henhouse door. Zsa Zsa is trying to horn in on Lucy's space, and Lucy is clucking mad. As Zsa Zsa blithely harangues her, Lucy is making a horrid noise that is closer to a banshee screeching than a chicken clucking. Being a solitary type myself, I feel bad for Lucy, so I go to the back room again and tap on the window. One-one thousand, two-one thousand, Zsa Zsa hears the tap and runs full speed out the henhouse door and to the coop door. Four beady eyes look up with as much hope as a chicken can summon. Sorry Girls, no snacks just yet.

A little while later, Lucy is clucking again. This time, the self-satisfied cluck of an egg well laid. I toggle to Chick TV and see that Zsa Zsa is now sitting in the favorite nest box. I'll tune in later for the results.

While making lunch and watching the last few minutes of the noon news, I hear excited clucking coming from you-know-where. I peer out the back window, careful to keep a low profile. Whoopee is still standing by the coop door, clucking and cawing. This is obviously a ploy to make me feel sorry for her and let her out. It works. I go out back and into the coop and open the back door to the hen compound. Behind the henhouse is the rest of my once-tawdry side yard, now home to these three urban chickens. The Girls tumble out behind me and immediately start to dig in the dirt and peck up bugs. Low, satisfied coo-clucks rise from the Girls' happy beaks. For now, they're happy. They are almost always happy.

Perhaps too happy. I had almost finished my work later that day when the unmistakably flustered high-speed clucking of an extremely disturbed chicken pierced the dull, late afternoon air.

Chicken alarm. Something's going on behind the coop in the run. I quickly dashed to the window overlooking the run, expecting to see a cat, a raccoon, something spooking the hens. Survey up; survey down. No cat, no 'coon, but something's not right. Something moves in the neighbor's yard. I look up, then blink twice to make sure my eyes are working correctly. Zsa Zsa is in the neighbor's backyard, strolling around the lawn, eating clover. How on earth did she traverse a 6-foot fence? She can't fly more than 3 feet in the air, and that's with a running start. Did she levitate? Did the other Girls give her a couple of legs up?

After retrieving my nomadic chicken and locking her and the others in the coop, I inspected the back run. Aha! Zsa Zsa didn't levitate, but Zsa Zsa did put her little bird brain to work. She realized — through contemplation or by sheer accident — that if she flew up onto the trash can (where I store her food), she just might be able to clear the fence from there. And so she did. I chuckled. The grass was, in that particular area, greener on the other side. And I moved the trash can.

I walk back into the henhouse. The Girls can't understand their nonnegotiable incarceration. They quickly forget about whatever their last thought was and crowd around me like groupies as I refill the food cylinder and replenish their water. Whoopee sneaks around my legs and steals a bite from the food tray. The other Girls follow behind her. I take my miniature bamboo rake and scoop together a small pile of droppings. The Girls eat on, nonplussed. I scoop the poop into the old dustbin I keep nearby and drop the droppings into the compost bin. Cycle of life and compost, right before my eyes.

After I toss around a little fresh straw, I check the nest boxes. What a surprise . . . all the Girls laid in the middle nest box today, even Whoopee! Wonder why they ignored their favorite box? Maybe Lucy didn't think the vibe was right after being heckled by Zsa Zsa. Perhaps the novelty of the newly claimed middle nest box stirred some desire in Whoopee to abandon her lackadaisical laying habits. No matter, I'm happy. Coincidentally, so are the chickens.

The late sun is making the coop hotter than it has been all day. I switch on the fan hanging in the corner and aim it at the Girls. At first, they scatter away, disturbed by the sudden breeze in their coop. By the time I leave the coop, eggs in hand, all three are lying in front of the fan, which gently blows about their feathers. For a moment, they look like three proper ladies on the deck of a fast-moving cruise ship, salt wind lifting up their scarves. The moment passes. Lucy stands up to scratch. Zsa Zsa gets a drink of water. And Whoopee jumps up after me, hoping to make a break for it as I exit the coop. No such luck. No matter. In a minute, the Girls will be happy all over again.

When I mention to other people that I have chickens, eyes and ears perk up, and I often find out about other flocks of city chickens nearby. Urban chicken keepers become noticeably excited during such impromptu "flock talks," and the people without chickens express first disbelief, then curiosity. Those without are amazed to hear of chickens living in populated cities and suburbs, and those with boast exuberantly to anyone who will listen about their chickens' eggs, antics, and easy care. If you're at the home of a chicken keeper, you'll all troop out to the coop so he or she can show off the birds and the beautful/funky/outrageous coop they live in. There are smiles, lots of smiles. City chicken keepers grin a lot, as if enjoying some private celebration.

I can't speak for others, but I know why I smile. Of course chickens are fun and provide fresh eggs. But there's something else. Since I've had chickens, a slow but important realization has seeped from the henhouse into my thick head. I realized that the world functions, and always has functioned, in pretty much the same way since time immemorial. The sun has always risen in the east. Rivers have always flowed down mountains. And chickens, with or without the "help" of human caretakers, have always laid eggs. I've realized that life is not as complex as we make it out to be. Rather, it's basically simple. My life — my interaction with the world — can be as simple as I want it to be.

Perhaps that's why other city chicken keepers are also smiling. With the help of their chickens, they've figured this out. Keeping a small flock of chickens, even two chickens, is an enjoyable and rewarding pastime that, incidentally, makes us a little more self-sufficient and brings us closer to the slow, measured beat of nature. Cluck. Cluck.

# Photo Gallery:
# A Chicken Extravaganza

CHAPTER 6 SUGGESTS SOME GOOD BREEDS for a backyard chicken flock. As is the case with other pets, picking chickens is a subjective experience. I based my selection of best backyard chickens on the breeds that lay lots of eggs but aren't broody, that have relatively calm demeanors, and that come in vivid colors. The chickens in the following photographs aren't the fanciest breeds but just a few of what I believe are the best breeds to keep in most small spaces as pets and for eggs.

Of course, you can't have chickens without a coop to keep them in. City chickens, like their country cousins, require basic shelter that is safe, warm, and dry. Many small flock owners build coops that are safe, warm, and dry, then toss in creative and whimsical architectural accessories that make their coops anything *but* basic. For the purposes of inspiration, you'll find here some photographs of a few of the coolest coops in the country. If you're like me, they'll both amuse and surprise, and they'll trigger new ideas for your own coop construction plans.

BREED
# Araucana

**Feather Colors:** Wide variety
**Eggshell Color:** Blue to army green
**Hen Body Weight:** 5 lb (2.3 kg)
**Comments:** Good layer.
Eggs smaller than other breeds listed.
Cautious and calm.

BREED
# Australorp

**Feather Colors:** Black, with a greenish tint

**Eggshell Color:** Dark brown

**Hen Body Weight:** 7 lb (3.2 kg)

**Comments:** Good layer. Cautious. Extra hardy in cold weather.

BREED
# Buff Orpington

**Feather Colors:** Buff camel
**Eggshell Color:** Medium brown
**Hen Body Weight:** 7 lb (3.2 kg)
**Comments:** Great layer. Friendly disposition. Hardy in cold weather.

BREED
# Leghorn

**Feather Colors:** White or dark red
**Eggshell Color:** White
**Hen Body Weight:** 5 lb (2.3 kg)
**Comments:** Great layer.
Friendly disposition.

BREED
# New Hampshire Red

**Feather Colors:** Chestnut red
**Eggshell Color:** Medium brown
**Hen Body Weight:** 6 lb (2.7 kg)
**Comments:** Great layer. Friendly and calm.

BREED
# Plymouth Rock

**Feather Colors:**
Barred black and white, buff and white

**Eggshell Color:** Light brown

**Hen Body Weight:** 6 lb (2.7kg)

**Comments:** Great layer.
Friendly and curious.

# Rhode Island Red

**Feather Colors:**
Dark red to reddish brown

**Eggshell Color:** Medium brown

**Hen Body Weight:** 6 lb (2.7 kg)

**Comments:** Cautious,
yet mellow. Great layer.

BREED

# Silver-Laced Wyandotte

**Feather Colors:** Contrasting silver-white and black

**Eggshell Color:** Medium brown

**Hen Body Weight:** 6 lb (2.7 kg)

**Comments:** Good layer. Attractive plumage pattern.

BREED
# Red Sex Link

**Feather Colors:** Dark red with black tail, wing feathers

**Eggshell Color:** Medium brown

**Hen Body Weight:** 5 lb (2.3 kg)

**Comments:** Good hybrid layer. Friendly disposition.

BREED
# Black Minorca

**Feather Colors:**
Black and white spotting
**Eggshell Color:** White
**Hen Body Weight:** 5 lb (2.3 kg)
**Comments:** Good layer
of white eggs.

# A Cool Coop Collection

The Lenham Poultry Range, a creation of Forsham Cottage Arks.

Marquee-style decor in a Seattle chicken run (owner John Bennet).

A veritable chicken villa (owners Ray and Robin Nichols).

The Henspa, a unique portable coop designed by Egganic Industries.

True country-coop style (courtesy of the imaginations of owners Jeff Stein and Margaret Kramer).

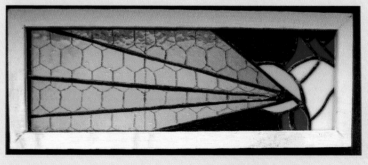

Stained-glass chick-chic window made by Brighton West, Atlanta, Georgia.

Two-story chicken chalet (owner Kimberly Minch).

Room with a view and her own flowerbox, too (owners Shelley Baker and Sonja Hunter).

# The Girls

The End.

# Appendix
■ ■ ■ ■ ■

## Summary of
## Selected City Municipal Codes

This listing will give you a good idea of chicken codes
across the country, current as of our publication date. Towns and
cities update their codes and ordinances on a regular basis, however,
so check in with your local health department to find the codes current to your area before you decide to take up chicken keeping.

# Summary of Selected City Municipal Codes

| State | City | City Code No. |
|---|---|---|
| Alabama | Mobile | Ch. 7, Art. IV, Div. 2, Sec. 7-102 and 7-103 |
| Alaska | Juneau | Part II, Title 8, Ch. 8 |
| Arizona | Phoenix | Ch. 8, Art. II, Sec. 8-7 and 8-8 |
| Arkansas | North Little Rock | Ch. 10, Art. IV, Sec. 10-166 thru 10-169 |
| California | Los Angeles | Ch. I, Art. 2, Sec. 12.05(A)(7)(a) |
| | San Diego | Ch. 4, Art. 2, Div. 7, Sec. 42.0708 & 42.0709 |
| | San Francisco | Health Code, Art. 1, Sec. 37 |
| | Santa Ana (Orange Co.) | Part II, Ch. 5, Art. I, Sec. 5-5.5 |
| | Anaheim | Title 8, Ch. 8, Sec. 08.070.010 |
| Colorado | Denver | Title II, Ch. 8, Art. IV, Div. 2, Sec. 8-91 |
| Connecticut | Hartford | Part II, Ch. 6, Sec. 6-4 |
| | New Haven | Title III, Ch. 7, Sec. 7-2 |
| Delaware | Wilmington | Part II, Ch. 3, Art. I, Sec. 3-15 |
| Florida | Miami | Part II, Ch. 6, Art. I, Sec. 6-1 |
| Georgia | Atlanta | Part II, Ch. 18, Art. II |
| Hawaii | Honolulu | Ch. 7, Sec. 7-2.5(d) |
| Idaho | Boise | Title 8, Sec. 8-08 |
| Illinois | Chicago | Title 7, Ch. 7-12, Sec. 7-12-300 |
| Indiana | Indianapolis | Title III, Ch. 531, Art. I, Sec. 531-102 |
| Iowa | Des Moines | Ch. 18, Sec. 18-4 |
| Kansas | Wichita (Sedgwick Co.) | Ch. 5, Art. I |
| | Topeka | Ch. 18, Art. VII, Sec. 18-291 |
| Kentucky | Louisville | Title 9, Ch. 90, Sec. 90.01 |
| | Lexington | Ch. 4, Sec. 4-10 |
| Louisiana | New Orleans (St. Charles Parish) | Part II, Ch. 4, Art. I, Sec. 4.3 |
| | Baton Rouge | Title 14, Ch. 2, Part III, Sec. 14.224 |

### Chicken Limits

25 chickens or fewer with permit; 200 feet from residences

No prohibition or restrictions on keeping chickens in the city

20 chickens or fewer with written permission of residences within 80 feet

Unspecified number of chickens with annual permit fee of $1.00; 75 feet from residences

5 chickens or fewer; no permit required

25 chickens or fewer with permit; 50 feet from residences

No permit for 4 or fewer chickens; 20 feet from any door or window

4 chickens or fewer; permit required for additional chickens

Unspecified number of chickens with annual permit fee

Unspecified number of chickens with annual permit fee of $50

Unspecified number of chickens; unlawful to keep poultry in area detrimental to public health

No prohibition on keeping chickens in the city; chickens can't roam at large or cause a nuisance

Chickens are prohibited

15 chickens or fewer, together with 30 or fewer baby chicks; permit required; 100 feet from residences

No prohibition on keeping chickens in the city; chickens can't roam at large

2 chickens or fewer; 300 feet from residences

No prohibition or restrictions on keeping chickens in the city

No prohibition on keeping chickens in the city; cannot kill chickens in the city

No prohibition or restrictions on keeping chickens in the city

25 chickens on the first acre of land, and 50 additional chickens on each acre thereafter

No prohibition or restrictions on keeping chickens in the city

Unspecified number of chickens; 50 feet from residences

No prohibition or restrictions on keeping chickens in the city

No prohibition on keeping chickens in the city; chickens can't roam at large

Unspecified number of chickens, with written permission of property owners within 300 feet

3 chickens or fewer; permit required for additional chickens; 10 feet from nearest property line, 50 feet from nearest residences

## Summary of Selected City Municipal Codes (Continued)

| State | City | City Code No. |
|---|---|---|
| Maine | Biddeford | Part II, Ch. 10, Art. I |
| Maryland | Baltimore (Howard Co.) | Title 17, Subdiv. 3, Sec. 17.300 |
| | Rockville | Ch. 3 |
| Massachusetts | Boston | Ch. 16, Sec. 16-1.8A |
| Michigan | Detroit | Part III, Ch. 6 |
| Minnesota | Minneapolis | Title 4, Ch. 70, Sec. 70.10 |
| Mississippi | Biloxi | Ch. 4, Art. I, Sec. 4-1-5 |
| Missouri | Springfield | Part II, Ch. 18, Art. I, Sec. 18-24 |
| Montana | Billings | Ch. 4, Art. 4-500, Sec. 4-501 |
| Nebraska | Omaha | Part II, Ch. 6, Div. 3, Sec. 6-233 |
| Nevada | Reno | Part II |
| New Hampshire | Nashua | Part II, Ch. 5, Art. I, Sec. 5-5.5 |
| | Concord | Title I, Ch. 13, Art. 13-1, Sec. 13-1-5 |
| New Jersey | Trenton | Health chapter |
| New Mexico | Las Cruces | Part II, Ch. 7, Art. 5, Sec. 7-227 |
| New York | Buffalo | Part II, Ch. 78, Art. I, Sec. 78-1 |
| North Carolina | Winston-Salem | Part III, Ch. 6, Sec. 6-6 |
| | Greensboro | Ch. 5, Art. I |
| North Dakota | Grand Forks | Part I, Ch. XI, Art. II, Sec. 11-0215 |
| Ohio | Toledo | Part 17, Title I, Ch. 1705, Sec. 1705.05 |
| Oklahoma | Oklahoma City | Ch. 8, Art. IV, Sec. 8-246 |
| Oregon | Portland | Title 13, Ch. 13.05 |
| Pennsylvania | Pittsburgh | Title 6, Art. III, Ch. 635, Sec. 635.02 |
| Rhode Island | Providence | Part II, Ch. 4, Art. I, Sec. 4-4 |
| South Carolina | Columbia | Ch. 4, Art. II, Sec. 4-33 |
| | Charleston | Ch. 5, Art. I, Sec. 5-9 |

**Chicken Limits**

| |
|---|
| No prohibition or restrictions on keeping chickens in the city |
| No prohibition or restrictions on keeping chickens in the city |
| No prohibition or restrictions on keeping chickens in the city |
| By permit for a fee of $20, and additional $10 fee for each 50 kept |
| No prohibition or restrictions on keeping chickens in the city |
| Unspecified number of chickens with annual permit fee of $10, plus written consent of 80 percent of residents within 100 feet |
| 2 chickens or fewer; no permit or fee; 150 feet from nearest property line |
| 12 chickens in a pen at least 144 sq. feet; up to 25 chickens with additional 12 sq. feet per each; 50 feet from nearest property line |
| No prohibition or restrictions on keeping chickens in the city |
| No prohibition on keeping chickens in the city; chickens can't roam at large |
| No prohibition or restrictions on keeping chickens in the city |
| No prohibition on keeping chickens in the city; 45 feet from residences |
| Unspecified number of chickens permitted; can't be a public nuisance or health hazard |
| Chickens are prohibited |
| Unspecified number of chickens permitted on half-acre parcel or larger |
| Unspecified number of chickens; no permit required; can't be detrimental to public health |
| Unspecified number of chickens; no permit or fee required; 20 feet from any property line, 50 feet from any principal residences |
| No prohibition or restrictions on keeping chickens in the city |
| Unspecified number of chickens permitted; can't be a public nuisance or health hazard |
| Unspecified number of chickens; permit required for all livestock within city limits |
| Unspecified number of hens permitted; 40 feet from residences |
| 3 chickens or fewer without a permit; more with permit; 50 feet from residences |
| No prohibition or restrictions on keeping chickens in the city |
| Chickens are prohibited |
| Chickens are prohibited |
| Unspecified number of chickens; permit required; 150 feet from residences |

# Summary of Selected City Municipal Codes (Continued)

| State | City | City Code No. |
|-------|------|---------------|
| South Dakota | Sioux Falls | Part II, Ch. 7, Art. I, Sec. 7-8 |
| Tennessee | Memphis | Part I, Ch. 5, Art. I, Sec. 5-3 |
| Texas | Dallas | Vol. I, Ch. 7, Art. I, Sec. 7-13.1, 7-15.1, 7-22 |
| | Houston | Ch. 6, Art. II, Sec. 6-31, 6-38 |
| | San Antonio | Part II, Ch. 5 |
| Utah | Salt Lake City | Title 8, Ch, 8, Sec. 8.08.010 |
| Vermont | Montpelier | Ch. 8, Art. I, Sec. 5–6 |
| Virginia | Richmond | Part II, Ch. 4, Art. I, Sec. 4-3 |
| Washington | Seattle | Title 23, Subtitle IV, Div. 2, Ch. 23.44, Sec. 23.44.048 |
| West Virginia | Beckley (near Charleston) | Part II, Ch. 3, Art. I, Sec. 3-10 |
| Wisconsin | Madison | Ch. 7, Sec. 7.29 |
| Wyoming | Jackson | Title 7, Ch. 7.04, Sec. 7.04.010 |
| | Laramie | Title 6, Ch. 6.06, Sec. 6.06.030 |

**Chicken Limits**

Unspecified number of chickens permitted; can't be a nuisance

No prohibition on keeping chickens in the city; chickens can't roam at large

Unspecified number of chickens; no permit required; no roosters; shelter required;
no disturbing noises or nuisance

7 chickens or fewer with permit only if person under doctor's orders for fresh chicken eggs

No prohibition or restrictions on keeping chickens in the city

Unspecified number of chickens with annual permit of $5 per hen, max. of $40 per year; 50
feet from residences

Unspecified number of chickens permitted; must be kept in an enclosure

Unspecified number of chickens subject to Health Dept. inspection

3 chickens or fewer without a permit on a 20,000 sq. foot lot


Unspecified number of chickens permitted; can't be a public nuisance or health hazard


Unspecified number of chickens permitted; can't be a public nuisance or health hazard

Chickens are prohibited

Unspecified number of chickens permitted; 20 feet from neighboring residences;
no roosters allowed

# Recommended Reading

*American Standard of Perfection* (latest edition), American Poultry Association. If you ever graduate from pet chickens to poultry fancy, this book is a must for your library.

*The Chicken Book*, by Page Smith and Charles Daniels (University of Georgia Press, 2000). A wide-ranging exploration of chickens from past to present.

*The Chicken Health Handbook*, by Gail Damerow (Storey Books, 1994). A complete guide to health care for and the epidemiology of chickens.

*The Chicken Tractor: The Permaculture Guide to Happy Hens and Healthy Soil*, by Andy Lee and Patricia Foreman (Good Earth Publications, 1998). Addressed more toward the poultry rancher than the backyard chicken keeper, but offers good information on making and using a portable chicken pen.

*Extraordinary Chickens*, by Stephen Green-Armytage (Harry N. Abrams, 2000). Lots of great photographs of some of the world's most unusual-looking chickens.

*The Fairest Fowl*, by Ira Glass and photographer Tamara Staples (Chronicle Books, 2001). Photographs of ribbon-winning chickens from poultry shows across the country. A good insider's look at the world of chicken fanciers.

*How to Build Small Barns and Outbuildings*, by Monte Burch (Storey Books, 1992). Detailed instruction on installing foundations and framing for small structures, like coops and henhouses.

*Storey's Guide to Raising Chickens*, by Gail Damerow (Storey Books, 1995). Lots of great technical information. A book for the more than just the casual chicken keeper.

# Chicks & Chicken Supplies

Contact the following retailers to purchase chicks (for pick-up or delivery) and all the supplies an urban chicken keeper might need.

**Cackle Hatchery**
P.O. Box 529
Lebanon, MO 65536
417-532-4581
www.cacklehatchery.com

**C. M. Estes Hatchery, Inc.**
P.O. Box 5776
Springfield, MO 65802
1-800-345-1420
www.esteshatchery.com

**Decorah Hatchery**
406 West Water Street
Decorah, IA 52101
563-382-4103
www.decorahhatchery.com

**Double-R Discount Supply**
3840 Minton Road
West Melbourne, FL 32904
866-325-7779
www.dblrsupply.com

**Egganic Industries**
3900 Milton Highway
Ringgold, VA 24586
800-783-6344
www.henspa.com
Makes the Henspa, a portable, ready-to-assemble henhouse.

**Forsham Cottage Arks**
16381 Black Run Road
Orange, VA 22960
540-672-6370
www.forshamusa.com
Makes a variety of ready-to-assemble henhouses.

**Grain-Belt Hatchery**
Box 125
Windsor, MO 65360
816-647-2711

**Heartland Hatchery**
RR 1, Box 177A
Amsterdam, MO 64723
660-267-3679
www.heartlandhatchery.com

**Hoffman Hatchery**
P.O. Box 129
Gratz, PA 17030
717-365-3694
www.hoffmanhatchery.com

**Hoover's Hatchery**
P.O. Box 200
Rudd, IA 50471
800-247-7014
www.hoovershatchery.com

**K & L Poultry Farm**
772 Morris Road
Aragon, GA 30104
706-291-1977
www.klpoultryfarm.com

**Marti Poultry Farm**
P.O. Box 27
Windsor, MO 65360
660-647-3156

**McKinney & Govero Poultry**
4717 Highway B
Park Hills, MO 63601
573-518-0535
or 573-431-4841
www.mckinneypoultry.com

**Murray McMurray Hatchery**
191 Closz Drive
P.O. Box 458
Webster City, IA 50595
800-456-3280
www.mcmurrayhatchery.com

**Moyer's Chicks**
266 East Paletown Road
Quakertown, PA 18951
215-536-3155

**Ridgway Hatcheries**
P.O. Box 306
LaRue, OH 43332
800-323-3825
www.ridgwayhatchery.com

**Smith Poultry & Game Bird Supplies**
14000 West 215th Street
Bucyrus, KS 66013
913-879-2587
www.poultrysupplies.com

**Stromberg's Chicks & Gamebirds Unlimited**
Box 400
Pine River, MN 56474
800-720-1134
www.strombergschickens.com

**Welp, Inc.**
P.O. Box 77
Bancroft, IA 50517
800-458-4473
www.welphatchery.com

# Chicken Information & Organizations

The following organizations and companies offer a range of chicken-related information.

**American Egg Board**
1460 Renaissance Drive
Park Ridge, IL 60068
www.aeb.org
Amusing industry site on the "incredible, edible" egg.

**The American Poultry Association**
133 Millville Street
Mendon, MA 01756
508-473-8769
www.ampltya.com
These are the folks who publish the *American Standard of Perfection,* the book on poultry breed standards.

**Backyard Chickens**
www.backyardchickens.com
A nicely laid-out site, with sections on coop design and construction, breeds, and a Chicken Chat Room to talk flock on-line.

**Feathersite**
www.feathersite.com
One of the best all-around poultry Web sites, with tons of information about chickens and other fowl. Breed index on-line with great color photos.

**Poultry Press**
P.O. Box 542
Connersville, IN 47331
765-827-0932
www.poultrypress.com
Publishes a newsletter promoting standard-breed poultry.

**Seattle Tilth Association**
4649 Sunnyside Avenue,
North, Room 1
Seattle, WA 98103
206-633-0451
www.seattletilth.org
A site that features organic gardening, urban ecology, composting, and recycling with links to the Seattle Tilth's Chicken Coop Tour photographs. The tour features nearly 20 urban coops open to the public for self-touring one day in summer.

# Index

■ ■ ■ ■ ■

Page numbers in **bold** indicate a table or box.

Page numbers in *italic* indicate a photo or illustration.